Springer Theses

Recognizing Outstanding Ph.D. Research

Aims and Scope

The series "Springer Theses" brings together a selection of the very best Ph.D. theses from around the world and across the physical sciences. Nominated and endorsed by two recognized specialists, each published volume has been selected for its scientific excellence and the high impact of its contents for the pertinent field of research. For greater accessibility to non-specialists, the published versions include an extended introduction, as well as a foreword by the student's supervisor explaining the special relevance of the work for the field. As a whole, the series will provide a valuable resource both for newcomers to the research fields described, and for other scientists seeking detailed background information on special questions. Finally, it provides an accredited documentation of the valuable contributions made by today's younger generation of scientists.

Theses are accepted into the series by invited nomination only and must fulfill all of the following criteria

- They must be written in good English.
- The topic should fall within the confines of Chemistry, Physics, EarthSciences, Engineering and related interdisciplinary fields such as Materials, Nanoscience, Chemical Engineering, Complex Systems and Biophysics.
- The work reported in the thesis must represent a significant scientific advance.
- If the thesis includes previously published material, permission to reproduce this must be gained from the respective copyright holder.
- They must have been examined and passed during the 12 months prior to nomination.
- Each thesis should include a foreword by the supervisor outlining the significance of its content.
- The theses should have a clearly defined structure including an introduction accessible to scientists not expert in that particular field.

More information about this series at http://www.springer.com/series/8790

Nasrin Nasrollahi

Improving Infrared-Based Precipitation Retrieval Algorithms Using Multi-Spectral Satellite Imagery

"Doctoral Thesis accepted by University of California, Irvine, USA"

Dissertation

Submitted in partial satisfaction of the requirements for the degree of Doctor of Philosophy in Civil Engineering

by

Nasrin Nasrollahi

Dissertation Committee:
Professor Soroosh Sorooshian, Chair
Professor Kuo-lin Hsu
Professor Brett Sanders
Professor Russell Detwiler
2013

Nasrin Nasrollahi
University of California, Irvine
Irvine
California
USA

ISSN 2190-5053 ISSN 2190-5061 (electronics)
ISBN 978-3-319-36332-5 ISBN 978-3-319-12081-2 (eBook)
DOI 10.1007/978-3-319-12081-2
Springer Cham Heidelberg New York Dordrecht London

Softcover reprint of the hardcover 1st edition 2015

Printed on acid-free paper

Springer is part of Springer Science+Business Media (www.springer.com)

To
Amir, my husband and my best friend
My Dad and Mom who shaped me into who I am
My sisters for their never ending love and
The joy of my life, Kian
Without their constantly support, encouragement and love, this dissertation would not have been possible.

Supervisor's Foreword

There is nothing more gratifying for a Professor than being asked to prepare a preface for the dissertation research of his student whose work is of the caliber worthy of being published in the Springer Theses Series. The dissertation research of Ms. Nasrin Nasrollahi at the University of California, Irvine exemplifies the best one can hope for from a doctoral student.

In a nutshell, her research dealt with the issue of how best one can use technological advances in observation systems and measure one of the key components of the global hydrologic cycle, namely precipitation, with the accuracy useful for various applications. The technology in this case is the availability of a variety of advanced instruments (infrared based channels, passive and active microwave radars, etc.) aboard a number of classes of environmental satellite systems (Geo Stationary, Polar Orbiting). Dr. Nasrollahi's contribution, which is the subject of this publication, is the integration of information from these multiple satellite sensors and multiple channels into the current precipitation estimation algorithms. In her work, she takes advantage of the recent NASA satellite CLOUDSAT which observes clouds and precipitation in high resolution and infuses that information into the current algorithms in order to eliminate some of the errors in existing data. In addition, she employs some of the recent machine-learning techniques to extract relevant information from large quantities of satellite data. Nasrin's final algorithm leads to a significant reduction in false rain signals, hence improving the quality of satellite-based estimates of precipitation.

One may ask "why is this important?" The answer lies in the fact that information about rainfall has become most important for two primary reasons. The first one is that changes in precipitation at the global scale hold clues about climate change with respect to its impact on the elements of the hydrologic cycle. Therefore having comprehensive estimates of precipitation in time and space covering the entire globe can give evidence about the shifting patterns of rainfall and how extreme events are changing. The second of course is how we as humans experience precipitation (rain and/or snow) in our daily lives. This could be simply knowing tomorrow's weather report i.e. if your area is getting rain or not or if you are going to expect flooding in your region. Such information about precipitation is therefore crucial for a range of applications such as dealing with hazards or improving the science and understanding the changes in the hydrologic cycle. Nasrin's dissertation is a research work contributing to this body of knowledge.

Department of Civil & Environmental Engineering, University of California, Irvine, CA, USA
7/31/2014

Soroosh Sorooshian

Preface

The Moderate Resolution Imaging Spectro-radiometer (MODIS) instrument aboard the NASA Earth Observing System (EOS) Aqua and Terra platform with 36 spectral bands provides valuable information about cloud microphysical characteristics and therefore precipitation retrievals. Additionally, CloudSat, selected as a NASA Earth Sciences Systems Pathfinder (ESSP) satellite mission, is equipped with a 94 GHz radar that can detect the occurrence of surface rainfall. The CloudSat radar flies in formation with Aqua with only an average of 60 s delay. The availability of surface rain occurrence based on CloudSat observation together with the multi-spectral capabilities of MODIS makes it possible to create a training data set to distinguish false rain areas based on their radiances in satellite precipitation products (e.g. Precipitation Estimation from Remotely Sensed Information using Artificial Neural Networks (PERSIANN). The brightness temperature of 6 MODIS water vapor and infrared channels are used in this study along with surface rain information from CloudSat to train an Artificial Neural Network model for no-rain recognition. The results suggest a significant improvement in detecting non-precipitating areas and reducing false identification of precipitation.

The second approach to identifying no-rain regions, developed in this study, is to find the areas covered with non-precipitating clouds. The cloud type data available from CloudSat is used as a target value to train an artificial neural network model to identify non-precipitating clouds such as cirrus and altostratus. Application of the trained model on two case studies investigated in this research, show significant improvements in near real-time PERSIANN rain estimations.

In addition, a cloud type classification algorithm was developed to classify clouds into seven different classes (cumulus (Cu), stratocumulus (Sc), altocumulus (Ac), altostratus (As), nimbostratus (Ns), high cloud and deep convective cloud). The classification model uses a self organizing features map to classify clouds based on multi-spectral MODIS data and CloudSat cloud types. The result of the classification model shows acceptable results for summertime. The winter season cloud classification is challenging due to dominance of low and middle level clouds. A better cloud classification algorithm for wintertime is achievable using active radar data and is beyond the capabilities of currently available remotely sensed multi-spectral information.

Parts of This Thesis Have Been Published in the Following Journal Articles:

Nasrollahi, N., Hsu K., Sorooshian S., 2013, An Artificial Neural Network model to reduce false alarms in satellite precipitation products using MODIS and CloudSat observations, Journal of Hydrometeorology, 14, 1872–1883.

AghaKouchak A., **Nasrollahi N**., Li J., Imam B., Sorooshian S., 2010, Geometrical Characterization of Precipitation Patterns, Journal of Hydrometeorology, 12 (2), 274–285, doi:10.1175/2010JHM1298.

Acknowledgements

I started my journey through graduate school at UCI in 2009. My time at UCI was not just another step on my educational path as I first thought, it has also been a great opportunity to get to know and work with some of the greatest people in my life and also in the science community. I cannot thank my advisor, Professor Soroosh Sorooshian enough for his great advise, motivation and inspiration throughout the years. He was not only an advisor in my research but also a mentor and a role model in my life. His international recognition, numerous awards and excellent work has been an inspiration. In addition, his great sense of humor made my time at UCI unforgettable. I would also like to thank my co-advisor, Professor Kuolin Hsu with whom I have worked very closely. His patience, guidance and passion for quality work have helped me tremendously. He always had great comments about my work and motivated me to work harder.

I would also like to thank my PhD committee members, Professor Brett Sanders and Professor Russell Detwiler, for their invaluable advice on this dissertation. I should also thank Diane Hohnbaum, Administrative Assistant in the Center for Hydrometeorology and Remote Sensing (CHRS), for her continuous support and help, and also Dan Braithwaite for his IT support and for providing technical help with the data.

I would also like to thank my friends at CHRS for their friendship, comments, support and encouragement. I had such an amazing time during our lunch discussions and power walks. I got to know them better and their positive energy enabled me to continue my work throughout the day. I would like to make a special mention of Sepideh Sarachi. She was not only a great lab-mate but also a great friend. I would also like to thank all my friends here in Irvine and elsewhere in the United States, who have been like a family to me.

My deepest gratitude goes to my family, my sisters and in-laws who are all the greatest support that I could have hoped for. I cannot thank them enough for being there for me all the time. Nasim, my twin sister, always understood my sadness and happiness even without any conversation. Azadeh brought laughter into my PhD life. Whenever I talked to Negar, I fed of her energy and passion. And Neda is the kindest sister one can imagine. I would also like to thank my parents. My father who always encouraged me to study to the highest level and my mother who is the kindest in the world and always supported me through the hard times.

Finally, I am so blessed to have my loving, supportive, encouraging, and patient husband Amir. From the early days of our friendship, engagement and then marriage he has been my inspiration and motivation for continuing to improve myself and my knowledge. Through all these years and now with the new member of our family, Kian, I feel that I have the happiest family in the world.

Primary financial support for this study was provided by NASA NESSF fellowship (NNX11AL33H), and US Army Research Office grant W911NF-11-1-0422.

Contents

1 Introduction to the Current State of Satellite Precipitation Products 1
1.1 The Importance of Precipitation in Water Resources 1
1.2 Precipitation Observation 2
1.3 Satellite-based Precipitation Estimation 2
1.4 Research Motivation 3
1.5 Objectives of this Dissertation 4
1.6 Dissertation Outline 4
References 5

2 False Alarm in Satellite Precipitation Data 7
References 11

3 Satellite Observations 13
3.1 MODIS 13
3.2 CloudSat 16
References 20

4 Reducing False Rain in Satellite Precipitation Products Using Cloudsat Cloud Classification Maps and Modis Multi-spectral Images 21
4.1 The Role of Multi-spectral Data in Satellite Precipitation Algorithms 21
4.2 Satellite Data 23
4.3 Methodology 23
4.4 Classification 26
4.5 Training Data Set 27
4.6 Application of the Model on Precipitation Events 28
4.7 Results and Discussions 28
References 30

5 Integration of CloudSat Precipitation Profile in Reduction of False Rain ... 33
5.1 Classification ... 33
5.2 Satellite Observations ... 33
5.3 Training Data Set ... 34
5.4 Application of the Model on Precipitation Events ... 35
5.5 Results and Discussions ... 35
5.6 Case Study ... 37
5.7 Conclusion ... 39
References ... 41

6 Cloud Classification and its Application in Reducing False Rain ... 43
6.1 Introduction ... 43
6.2 MODIS Cloud Mask ... 44
6.3 Image Classification Using Self Organizing Maps ... 44
6.4 Data Pre-processing ... 46
6.5 Training and Validation Datasets, Summer Season ... 47
6.6 Training and Validation Datasets, the Winter Season ... 48
6.7 SOFM Model for Summer Season ... 52
6.8 Validation of Cloud Classification Model, Summertime ... 55
6.9 SOFM Model for the Winter Season ... 56
6.10 Validation of Cloud Classification Model, the Winter Season ... 61
6.11 Conclusion ... 61
References ... 62

7 Summary and Conclusions ... 65
7.1 Future Work ... 67

List of Figures

Fig. 1.1 Vertical structure of clouds and corresponding IR brightness temperature 3

Fig. 2.1 The definition of Probability Of Detection (POD) and False Alarm Ratio (FAR) 8

Fig. 2.2 False Alarm Ratio (*FAR; top three panels*) and Probability of Detection (*POD; bottom three panels*) of the PERSIANN precipitation data over the contiguous US (from 2005 to 2008). **a, d** 4 years results. **b, e** Summer season results. **c, f** Winter season results 9

Fig. 2.3 False Alarm Ratio (*FAR; top three panels*) and Probability of Detection (*POD; bottom three panels*) of the TRMM 3B42 precipitation data over the contiguous US (from 2005 to 2008). **a, d** 4 years results. **b, e** Summer season results. **c, f** Winter season results 10

Fig. 2.4 The TRMM, PERSIANN and stage II precipitation pattern for rainfall rates above the 50th (**a–c**), 75th (**d–f**), 90th (**g–i**) percentiles.... 11

Fig. 3.1 Atmospheric windows in the electromagnetic spectrum. (Source: Jensen 2007) 14

Fig. 3.2 The A-Train constellation. (Source: http://cloudsat.atmos.colostate.edu/education/satellites) 17

Fig. 3.3 Common types of clouds. (Source: http://airlineworld.files.wordpress.com/2008/07/cloud_types.gif) 18

Fig. 4.1 Part a. An example of the CloudSat cloud classification map and MODIS brightness temperature data on August 13th, 2008 (05:45 UTC). **a** Track of CloudSat passing through a storm measured by Stage IV precipitation data. **b** CloudSat vertical cloud profile. **c** PERSIANN precipitation data (mm/hr). **d** Stage IV precipitation data (mm/hr). **e** MODIS brightness temperature at 11 μm (*Kelvin*). Part b. **f** MODIS BTD[8.5–11] (*Kelvin*). **g** Radar reflectivity (*dBZ*) 24

Fig. 4.2 False rain identification algorithm using cloud classification data) ... 25
Fig. 4.3 Schematic of the feed-forward three-layer perceptron with 6 input variables. The final output layer provides the rain/no rain detection) ... 26
Fig. 4.4 False alarm detection using MODIS 6 spectral bands and CloudSat CLD-CLASS August 13, 2008 (0545 UTC)) ... 29
Fig. 4.5 False alarm detection using MODIS 6 spectral bands and CloudSat CLD-CLASS, July 22, 2008 (0805 UTC)) ... 30

Fig. 5.1 The ANN model weights for summer (*left*) and winter (*right*) seasons ... 34
Fig. 5.2 Distribution of different cloud types in the case of correct NR detection (*top*) and misclassifications (*bottom*), for the summer of 2007 ... 36
Fig. 5.3 Distribution of different cloud types in the case of correct NR detection (*top*) and misclassifications (*bottom*), for the winter of 2007 ... 37
Fig. 5.4 Performance of the ANN model in identifying false rain locations on August 5th, 2007 (05:00 UTC). **a** Stage IV precipitation data (mm/h). **b** PERSIANN data (mm/h). **c** Model performance in FAR detection ... 39
Fig. 5.5 Performance of the ANN model on November 6th, 2007 (03:00 UTC). **a** Stage IV precipitation data (mm/h). **b** PERSIANN data (mm/h). **c** Model performance in FAR detection ... 40

Fig. 6.1 Structure of a SOFM model ... 45
Fig. 6.2 A schematic diagram of the SOFM training (Source: http://en.wikipedia.org/wiki/Self-organizing_map) ... 46
Fig. 6.3 A box-plot of 6 MODIS channel input data, summer 2008 ... 48
Fig. 6.4 Distribution of different cloud types in the input data, summer 2008 ... 49
Fig. 6.5 Distribution of different cloud types in the validation dataset, summer 2007 ... 49
Fig. 6.6 Distribution of different cloud types in the training dataset, winter 2010 ... 50
Fig. 6.7 Distribution of different cloud types in the validation dataset, winter 2007 ... 50
Fig. 6.8 A box-plot of 7 MODIS channel input data, winter 2010 ... 51
Fig. 6.9 Distribution of different samples on a SOFM map ... 52
Fig. 6.10 Weight of different input features on the SOFM map ... 53
Fig. 6.11 Number of cloud samples on each SOFM cluster ... 54
Fig. 6.12 Probability of various cloud types on SOFM clusters ... 54
Fig. 6.13 Decision matrix, summer season ... 55
Fig. 6.14 Decision confidence of summer classification ... 56

Fig. 6.15 The accuracy of summertime cloud classification 57
Fig. 6.16 Distribution of training samples on the SOFM map 57
Fig. 6.17 Weight of different input features on the SOFM map....................... 58
Fig. 6.18 Number of cloud samples on each SOFM cluster 58
Fig. 6.19 Probability of various cloud types on SOFM clusters 59
Fig. 6.20 Decision matrix, the winter season 60
Fig. 6.21 Decision confidence of winter classification 60
Fig. 6.22 The accuracy of winter season cloud classification.......................... 61

Fig. 7.1 Availability of CloudSat, MODIS and future GOES-R satellites....... 68

List of Tables

Table 3.1 Information about 36 spectral channels of MODIS instrument........ 15

Table 3.2 Different cloud types characteristics. (Source: Wang and Sassen (2007))........ 19

Table 5.1 NR misclassification error percentage for summer and winter seasons........ 38

Table 5.2 Model performance presented in Figs. 5.4 and 5.5........ 40

Table 6.1 Initial classes from MODIS cloud mask........ 45

Chapter 1
Introduction to the Current State of Satellite Precipitation Products

1.1 The Importance of Precipitation in Water Resources

Floods cause more deaths than any other natural disaster around the world, with a death toll of more than 5000 individuals per year. The United States also has a long history of catastrophic flooding. Most flood deaths are due to flash floods that occur within a few minutes or hours of excessive rainfall over a region. Flash floods cause more deaths annually than any other weather phenomenon in the United States, with a death toll of more than 1000 individuals over a 10-year period between 1983–1992, and an average of greater than $ 2 billion in annual losses over the same 10-year period (U.S. Army Corps of Engineers 1993). In the period between 1950 and 1997, the National Weather Service reported an average of 110 deaths per year in flood-related accidents. In addition, recent major flood events around the world (e.g. Aug 2010 severe flood in Pakistan, Dec 2010 flood in Brazil) emphasize the need for hydro-meteorological information to address natural hazards with major socio-economic impacts.

The two key elements that contribute to flash floods are rainfall intensity and duration. Other factors such as soil moisture, topography and land cover also play an important role (Song et al. 2014). Land use change due to urbanization is also a factor that increases the risk of flooding. In the US, urbanization has increased the magnitude of floods during the twentieth century and many urban watersheds suffered from greater floods (Hollis 1975). In addition, human-induced climate change has a direct impact on precipitation. Increase in the water holding capacity of the atmosphere due to a change in atmospheric temperature leads to increased water vapor in the atmosphere. Hence, in the future more intense precipitation events will be observed which will increase the risk of flooding (Trenberth 2011).

Because of the nature of flash floods, reliable estimation of precipitation is important to predict and manage water resources, hazard preparedness and climate studies (Ajami et al. 2008; AghaKouchak and Nakhjiri 2012; Anderson et al. 2008; Hao et al. 2014; Damberg and AghaKouchak et al. 2014; Tabari et al. 2014). Availability of real-time rainfall data plays a major role in prediction of floods and affects decision making.

N. Nasrollahi, *Improving Infrared-Based Precipitation Retrieval Algorithms Using Multi-Spectral Satellite Imagery,* Springer Theses, DOI 10.1007/978-3-319-12081-2_1

1.2 Precipitation Observation

Rain gauge estimation is the traditional method of precipitation measurement. However, spatial and temporal variability of precipitation makes it difficult to rely on gauge point measurements. Gauge distribution is uneven around the world and usually depends on the population of the area. Needless to mention, there are no gauges over the oceans, insufficient to capture regional precipitation variability. Radars, on the other hand, provide high resolution estimates of precipitation. However, radar networks are not available everywhere in the world. In addition, radar coverage area becomes smaller at lower altitudes (e.g., 1000 m above the ground level) in comparison to higher elevations (e.g., 3000 m above ground level), mainly due to blockage problem in the mountainous regions (e.g. western United States, Maddox et al. 2002). Therefore, an alternate method to estimate precipitation globally with high spatial and temporal resolution and reliable accuracy is needed. Using satellite remote sensing technology helps to derive a better global coverage of precipitation. Satellites observe the Earth from the space and are able to gather some information that cannot be made available from ground based instruments. The National Aeronautics and Space Administration (NASA), National Oceanic and Atmospheric Administration (NOAA), the European Organization for the Exploitation of Meteorological Satellites (EUMETSAT) and many other internationally sponsored satellite missions have provided valuable information that can be used to estimate precipitation. Global precipitation data can be utilized in disaster management and decision making operations.

1.3 Satellite-based Precipitation Estimation

The main sensors to estimate precipitation from space are visible (VIS), Infrared (IR) and Passive Microwave (PMW). VIS and IR data are available from Geostationary Earth Orbiting (GEO) and Low-Earth orbiting (LEO) satellites. However, VIS and IR channels do not measure precipitation directly. Instead, they measure cloud albedo and cloud top temperature that can be associated with the precipitation rate using an indirect relationship. One limitation of IR-based algorithms is that non-precipitating cold clouds at high altitudes are often falsely identified as precipitating clouds, resulting in false precipitation estimates. Intense precipitation is correlated with cold clouds. However, the converse relationship may not be true (Fig. 1.1). In addition to this issue, orographically induced precipitation or precipitating warm clouds (e.g. stratiform) may cause precipitation, which is not easily identified with current algorithms (Joyce et al. 2004). The misclassification of rain/no-rain clouds is one of the major issues facing IR-based algorithms (Arkin and Xie 1994; Turk and Miller 2005; Behrangi et al. 2009). Adding information about visible channels helps to improve rain estimation however, VIS data are not available during the night time.

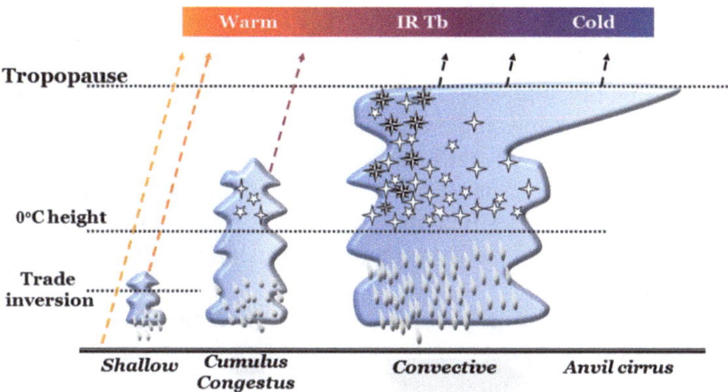

Fig. 1.1 Vertical structure of clouds and corresponding IR brightness temperature

In addition to IR, VIS and water vapor channels, LEO satellites are equipped with passive microwave (PMW) sensors that measure the thermal emission and scattering of raindrops. PMW remote sensing of precipitation is recognized as a more reliable source of precipitation estimation from space (Adler et al. 2001; Ebert et al. 1996). However, LEO satellites have low temporal resolution of only one or two times a day for a specific location on earth (Marzano et al. 2004). Many LEOs are orbiting the Earth, therefore PMW data from LEOs are operationally available every few hours. To date, PMW sensors are not carried on GEO satellites because of technical challenges (Joyce et al. 2004). In addition to their low temporal and spatial resolution, PMW sensors are more reliable over oceans because of the complexity of land surface emissivity.

1.4 Research Motivation

Reliable estimation of precipitation is important to predict and manage water resources. However, spatial and temporal variability of precipitation makes it difficult to rely on sparse gauge point measurements for remote regions. Higher spatial and temporal resolutions of satellite observations are the main advantages of remotely sensed precipitation estimates over in-situ measurements. Since they are an indirect method to estimate precipitation, they are also associated with uncertainties. Reducing false rain in IR-based precipitation algorithm will improve the quality of satellite estimations significantly.

CloudSat radar has the ability to provide a 3D structure of clouds from space. CloudSat data can be used to add additional information to the precipitation algorithm and to cloud detection. This additional source of information will improve the quality of rain estimation algorithms.

1.5 Objectives of this Dissertation

In this dissertation, the application of multi-spectral data and statistical classification techniques in improving single channel IR precipitation algorithms is explored. Multi-spectral data available from Moderate resolution Imaging Spectroradiometer (MODIS) images and CloudSat are two sources of information that are used to improve the quality of rain estimations and reduce false rain detection. CloudSat data is used to train a Neural Network model using MODIS data as input to identify false rain locations.

Application of CloudSat data in cloud classification model is also investigated. The cloud classification model can be used to find precipitating clouds and run the precipitation algorithm only on those clouds.

The objectives of this dissertation are:

1. ***Using multi-spectral data in satellite precipitation algorithms will help improve precipitation algorithms. There is a need to move from single IR channel estimations to multi-channel precipitation algorithms. The first objective of this dissertation is to show the effectiveness of using multi-spectral data in satellite precipitation estimation.***
2. ***The second objective of this dissertation is to show that satellite precipitation algorithms will benefit from information on cloud structure and characteristics. Clouds create precipitation, and adding information about different types of clouds will improve precipitation algorithms.***
3. ***The main reason for false rain observations in satellite-based products is the presence of high cirrus clouds. These highly elevated clouds have cold cloud tops in IR imagery. Therefore, they show false rain signals in satellite-based estimations. The third objective is to show that by identifying and filtering cold cirrus clouds false rain reduces.***

The answer to the above mentioned questions will be addressed in this dissertation.

1.6 Dissertation Outline

This dissertation is organized into six sections: Chapter 2 explains false alarm in satellite precipitation and how we can identify false rain. Chapter 3 is devoted to explaining satellite observations. Chapter 4 is about reducing false rain using CloudSat cloud classification data and Chap. 5 examines false rain reduction using CloudSat surface precipitation presence dataset. A cloud classification algorithm is presented in Chap. 6 and the summary and future works are described in Chap. 7.

References

Adler, R., Kidd, C., Petty, G., Morrissey, M., & Goodman, H. (2001). Intercomparison of global precipitation products: The third precipitation intercomparison project (pip-3). *Bulletin of the American Meteorological Society, 82,* 1377–1396.

AghaKouchak, A., & Nakhjiri, N. (2012). A near real-time satellite-based global drought climate data record. *Environmental Research Letters, 7*(4), 044037.

Ajami, N., Hornberger, J., & Sunding, D. (2008). Sustainable water resource management under hydrological uncertainty. *Water Resources Research, 44,* W11406.

Anderson, J., Chung, F., Anderson, M., Brekke, L., Easton, D., Ejeta, M., Peterson, R., & Snyder, R. (2008). Progress on incorporating climate change into management of California's water resources. *Climatic Change, 87*(1), 91–108.

Arkin, P., & Xie, P. (1994). The global precipitation climatology project—1st algorithm intercomparison project. *Bulletin of the American Meteorological Society, 75,* 401–419.

Behrangi, A., Hsu, K., Imam, B., Sorooshian, S., & Kuligowski, R. (2009). Evaluating the utility of multi-spectral information in delineating the areal extent of precipitation. *Journal of Hydrometeorology, 10*(3), 684–700.

Damberg, L., & AghaKouchak, A. (2014). Global Trends and Patterns of Droughts from Space, Theoretical and Applied Climatology, *117*(3), 441–448, doi:10.1007/s00704-013-1019-5.

Ebert, E., Manton, M., Arkin, P., Allam, R., Holpin, C., & Gruber, A. (1996). Results from the GPCP algorithm intercomparison programme. *Bulletin of the American Meteorological Society, 77,* 2875–2887.

Hao, Z., AghaKouchak, A., Nakhjiri, N., & Farahmand, A. (2014). Global Integrated Drought Monitoring and Prediction System, Scientific Data, 1:140001, 1–10, doi:10.1038/sdata.2014.1.

Hollis, G. E. (1975) Effect of urbanization on floods of different recurrence interval. *Water Resources Research, 11*(3), 431–435.

Joyce, R., Janowiak, J., Arkin, P., & Xie, P. (2004). CMORPH: A method that produces global precipitation estimates from passive microwave and infrared data at high spatial and temporal resolution. *Journal of Hydrometeorology, 5,* 487–503.

Maddox, R. A., Zhang, J., Gourley, J. J., & Howard, K. W. (2002). Weather radar coverage over the contiguous United States. *Weather and Forecasting, 17*(4), 927–934.

Marzano, M., Palmacci, F. S., Cimini, D., Giuliani, G., & Turk, F. (2004). Multivariate statistical integration of satellite infrared and microwave radiometric measurements for rainfall retrieval at the geostationary scale. *IEEE Transactions on Geoscience and Remote Sensing, 42,* 1018–1032.

Song, X., Zhang, J., AghaKouchak, A., Sen Roy, S., Xuan, Y., Wang, G., He, R., Wang, X., Liu, C. (2014). Rapid urbanization and changes in trends and spatio-temporal characteristics of precipitation in the Beijing metropolitan area, Journal of Geophysical Research, doi:10.1002/2014JD022084.

Tabari, H., AghaKouchak, A., Willems, P. (2014). A perturbation approach for assessing trends in precipitation extremes, Journal of Hydrology, doi:10.1016/j.jhydrol.2014.09.019.

Trenberth, K. E. (2011). Changes in precipitation with climate change. *Climate Research, 47*(1–2), 123–138.

Turk, F., & Miller, S. (2005). Toward improving estimates of remotely sensed precipitation with MODIS/AMSR-E blended data techniques. *IEEE Transactions on Geoscience and Remote Sensing, 43,* 1059–1069.

Chapter 2
False Alarm in Satellite Precipitation Data

Evaluation of satellite precipitation algorithms is essential for future algorithm development. This is why many previous studies are devoted to the validation of satellite-based observations (e.g., Tian et al. 2009; Amitai et al. 2009; AghaKouchak et al. 2010b; Zhou 2008; Gochis et al. 2009; Yilmaz et al. 2005; Shen et al. 2010; Dinku et al. 2008; AghaKouchak et al. 2009; Liu et al. 2009; Sapiano and Arkin 2009; AghaKouchak et al. 2012). For instance, Tian et al. (2009) analyzed the error of six high-resolution satellite products versus a gauge-based estimate, and reported regional and seasonal variations of error patterns in the contiguous US. They conclude that satellite products tend to overestimate rainfall in the summer and underestimate it in the winter. Sapiano and Arkin (2009) also confirmed that satellites overestimate summertime convective storms over the US. Using Volumetric False Alarm Ratio, AghaKouchak et al. (2011) showed that several satellite products exhibit high false alarm rate for rainfall, especially at high quantiles of observation.

To investigate false alarms in satellite-based precipitation products, we conducted a validation study to compare PERSIANN and TRMM TB42 precipitation data with ground based measurements. False Alarm Ratio (FAR) and Probability of Detection (POD) are calculated for the time period between 2005 and 2008 over the US. The FAR is the ratio of falsely identified rainy pixels to the total number of rainy pixels in satellite data, whereas the POD measures the fraction of observed precipitation that was correctly forecasted (the ratio of the total number of times that rainfall was correctly forecasted to the total number of observed rainy pixels (Wilks 2006)). Figure 2.1 explains the definition of POD and FAR.

In the current study, the Stage IV radar-based multi-sensor precipitation estimates (MPE), available from the National Center for Environmental Prediction (NCEP), are used as the reference data. The Stage IV precipitation data are adjusted for various biases using rain gauge measurements (Lin and Mitchell 2005) and are considered the best area approximation among the currently available area-average rainfall datasets (AghaKouchak et al. 2010a; AghaKouchak et al. 2010c, d). Stage IV data is aggregated into 0.25 degree spatial and 3 hourly temporal resolutions, which is the same as the PERSIANN precipitation data. Figure 2.2 shows the FAR and POD for: (a) the entire period of 4 years, (b) the summer and (c) the winter

N. Nasrollahi, *Improving Infrared-Based Precipitation Retrieval Algorithms Using Multi-Spectral Satellite Imagery,* Springer Theses, DOI 10.1007/978-3-319-12081-2_2

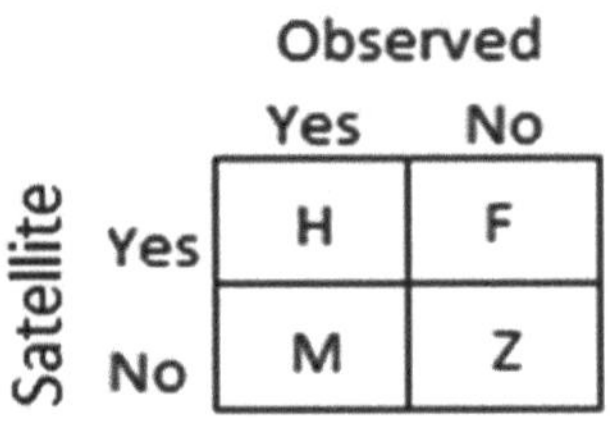

Fig. 2.1 The definition of Probability Of Detection (POD) and False Alarm Ratio (FAR)

		Observed	
		Yes	No
Satellite	Yes	H	F
	No	M	Z

seasons for the PERSIANN precipitation product (precipitation threshold is considered as 0.05 mm/h). Figure 2.2 reveals very high FARs over the central and western US and a lower FAR over the eastern US on average. Higher FAR is associated with presence of high cirrus clouds, especially in the winter. As discussed by Tian et al. (2009), PERSIANN data demonstrates higher FAR over the western US in the winter. The average POD is about 60 % over the central US and very low over the southwestern region. Low POD on the eastern and western side of the continent is associated with missed precipitation over these regions.

The missed precipitation may be caused by snow cover on the ground at higher latitudes or over the Rockies, and by the inability to catch warm rain processes or short-lived convective storms at lower latitudes, or maritime precipitation along the west coast (Tian et al. 2009).

Generally, probability of detection of satellite precipitation seems to be better during the summer seasons, perhaps due to a dominance of convective storms. On the other hand, the FAR is very high during the wintertime because of the presence of non-precipitating, high, cold clouds. Additionally, the presence of snow and ice on the ground and the inability of Passive Microwave (PMW) sensors to measure snowfall over snow or ice covered surfaces increase the error in satellite precipitation estimations and result in higher FARs during the wintertime. Figure 2.3 shows the same results for Tropical Rainfall Measuring Mission (TRMM) Multi-satellite Precipitation Analysis (TMPA) 3B42 precipitation data. Overall, both datasets show higher false rain during the winter and better estimations during summertime. Finally, it is worth mentioning that radar coverage is limited over the western region of the US (with the very high false alarm shown in Fig. 2.2 because of beam blockage in mountainous terrain). The Stage IV data has a large number of missing data over the Pacific Northwest region therefore, the precipitation data for this region is not included in the analysis.

In addition to calculation pixel-based false alarm, object-based approaches also show differences in precipitation estimations. For example, comparing different satellite-based precipitation patterns with the stage II radar-based precipitation pattern shows how the satellite estimations differ from radar observations. Figure 2.4 shows two satellite images that occurred at 0900 UTC 24 September 2005 during Hurricane Rita, with the spatial and temporal resolutions of 0.25×0.25 and 3 h [a: Tropical Rainfall Measuring Mission (TRMM) 3B42, Huffman et al. (2007); b: PERSIANN, Sorooshian et al. (2000)]. Hurricane Rita was one of the most intense tropical cyclones that made landfall on the U.S. Gulf Coast. Notice that in panels a to c, only precipitation values above the 50th percentile threshold are considered

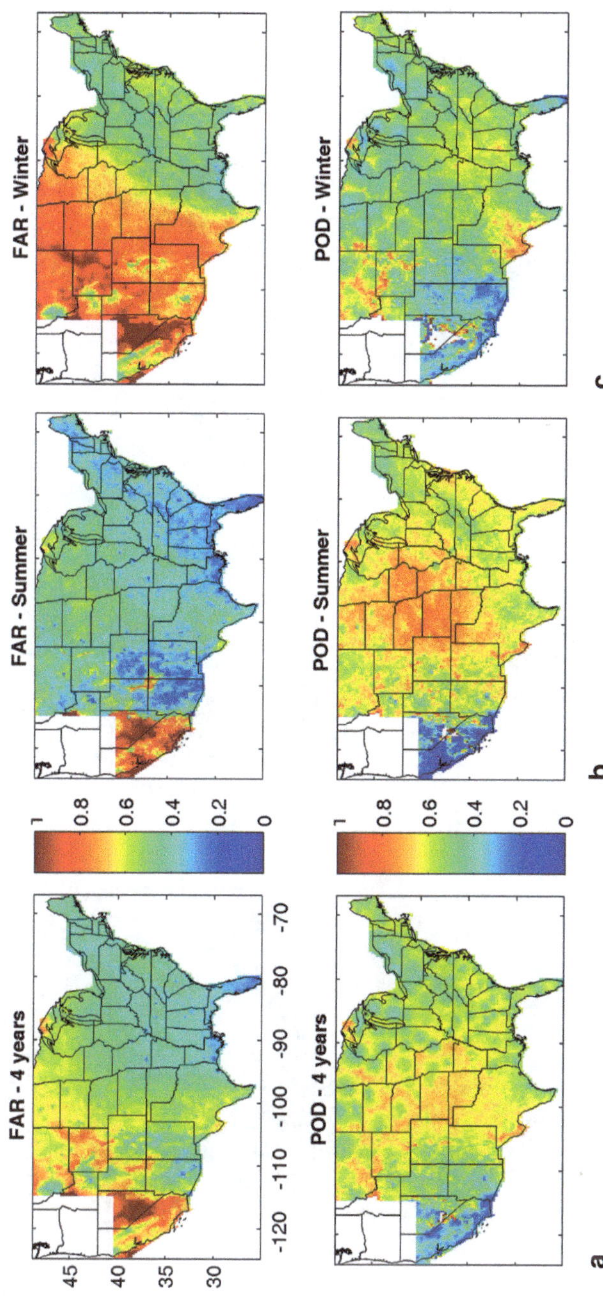

Fig. 2.2 False Alarm Ratio (*FAR; top three panels*) and Probability of Detection (*POD; bottom three panels*) of the PERSIANN precipitation data over the contiguous US (from 2005 to 2008). **a, d** 4 years results. **b, e** Summer season results. **c, f** Winter season results

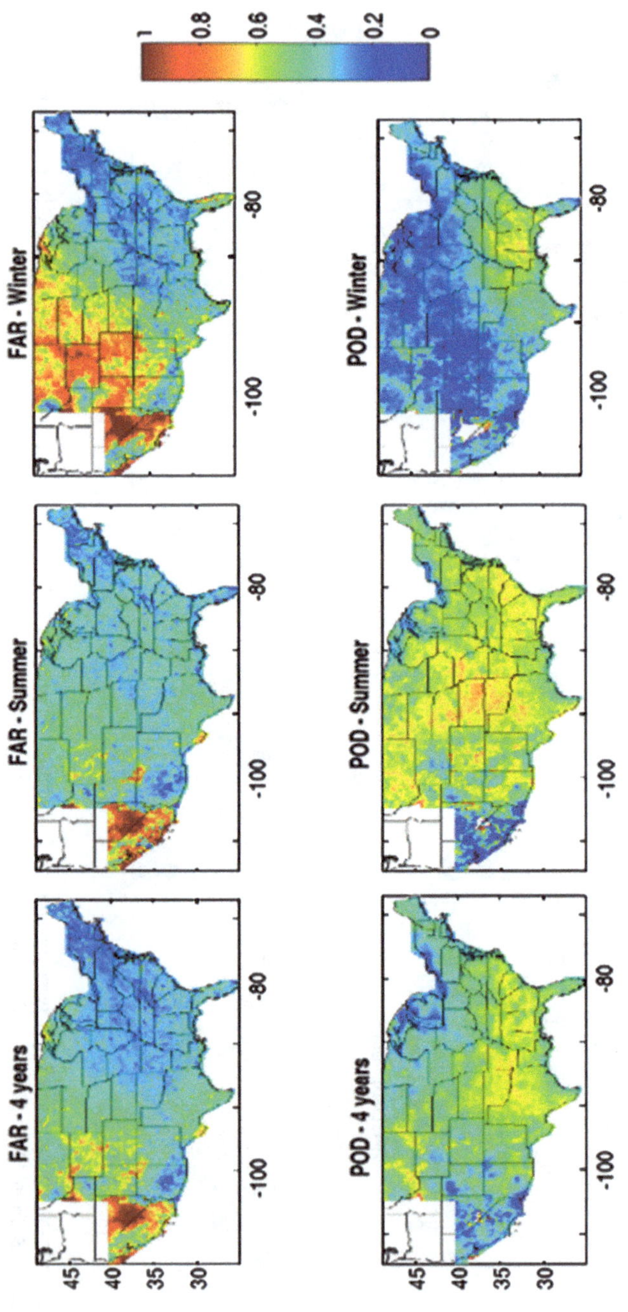

Fig. 2.3 False Alarm Ratio (*FAR; top three panels*) and Probability of Detection (*POD; bottom three panels*) of the TRMM 3B42 precipitation data over the contiguous US (from 2005 to 2008). **a, d** 4 years results. **b, e** Summer season results. **c, f** Winter season results

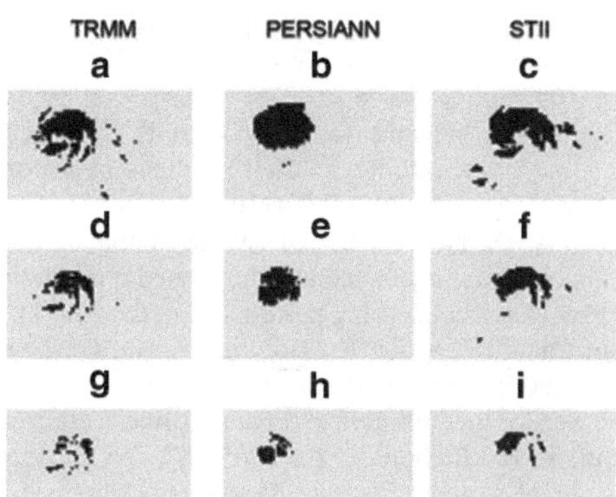

Fig. 2.4 The TRMM, PERSIANN and stage II precipitation pattern for rainfall rates above the 50th (**a–c**), 75th (**d–f**), 90th (**g–i**) percentiles

to avoid small rainfall rates. Panel c displays the corresponding stage II image. The stage II data provide estimates of precipitation using a combination of radar and rain gauge measurements. The data is available on the Hydrologic Rainfall Analysis Project (HRAP) grid, with a spatial resolution of approximately 4 km. The stage II data are aggregated in space to match the spatial resolution of TRMM and PERSIANN data. Panels d–f and g–i present similar figures for precipitation values exceeding the 75th and 90th percentiles, respectively. The domain of all figures includes 94×47 pixels, each being 0.25×0.25. With respect to the shape of patterns, the TRMM seem to be closer to the stage II data. However, for a higher threshold of 75th and 90th percentiles, the pattern of PERSIANN precipitation is more similar to those of stage II. It should be noted that the above example is provided to show differences in the pattern of precipitation and should not be considered as validation of satellite precipitation data.

References

AghaKouchak, A., Nasrollahi, N., & Habib, E. (2009). Accounting for Uncertainties of the TRMM Satellite Estimates, Remote Sensing, *1*(3), 606–619, doi:10.3390/rs1030606.

AghaKouchak, A., Bardossy, A., & Habib, E. (2010a). Copula-based uncertainty modeling: Application to multi-sensor precipitation estimates. *Hydrological Processes, 24*(15), 2111–2124.

AghaKouchak, A., Nasrollahi, N., Li, J., Imam, B., & Sorooshian, S. (2010b). Geometrical characterization of precipitation patterns. *Journal of Hydrometeorology, 12*(2), 274–285.

AghaKouchak, A., Habib, E., & Bardossy, A. (2010c). Modeling Radar Rainfall Estimation Uncertainties: Random Error Model, Journal of Hydrologic Engineering, *15*(4), 265–274, doi:10.1061/(ASCE)HE.1943-5584.0000185.

AghaKouchak, A., Bardossy, A., & Habib, E., (2010d). Conditional Simulation of Remotely Sensed Rainfall Fields Using a Non-Gaussian V-Transformed Copula, Advances in Water resources, *33*(6), 624–634, doi:10.1016/j.advwatres.2010.02.010.

AghaKouchak, A., Behrangi, A., Sorooshian, S., Hsu, K., & Amitai, E. (2011). Evaluation of satellite-retrieved extreme precipitation rates across the central United States. *Journal of Geophysical Research, 116,* D02115.

AghaKouchak, A., Mehran, A., Norouzi, H., & Behrangi, A. (2012). Systematic and random error components in satellite precipitation data sets. *Geophysical Research Letters, 39,* L09406.

Amitai, E., Llort, X., & Sempere-Torres, D. (2009). Comparison of TRMM radar rainfall estimates with NOAA next-generation QPE. *Journal of the Meteorological Society of Japan, 87,* 109–118.

Dinku, T., Chidzambwa, S., Ceccato, P., Connor, S., & Ropelewski, C. (2008). Validation of high-resolution satellite rainfall products over complex terrain. *International Journal of Remote Sensing, 29*(14), 4097–4110.

Gochis, D., Nesbitt, S., Yu, W., & Williams, S. (2009). Comparison of gauge-corrected versus non-gauge corrected satellite-based quantitative precipitation estimates during the 2004 name enhanced observing period. *Atmosfera, 22*(1), 69–98.

Huffman, G., Adler, R., Bolvin, D., Gu, G., Nelkin, E., Bowman, K., Stocker, E., & Wolff, D. (2007). The TRMM multi-satellite precipitation analysis: Quasi-global, multiyear, combined-sensor precipitation estimates at fine scale. *Journal of Hydrometeorology, 8,* 38–55.

Lin, Y., & Mitchell, K. E. (2005). The NCEP stage II/IV hourly precipitation analyses: Development and applications. Preprints, 19th Conf. on Hydrology, San Diego, CA, Am. Meteorol. Soc., 1.2.

Liu, Z., Rui, H., Teng, W., Chiu, L., Leptoukh, G., & Kempler, S. (2009). Developing an online information system prototype for global satellite precipitation algorithm validation and intercomparison. *Journal of Applied Meteorology and Climatology, 48*(12), 2581–2589.

Sapiano, M., & Arkin, P. (2009). An intercomparison and validation of high-resolution satellite precipitation estimates with 3-hourly gauge data. *Journal of Hydrometeorology, 10*(1), 149–166.

Shen, Y., Xiong, A., Wang, Y., & Xie, P. (2010). Performance of high-resolution satellite precipitation products over China. *Journal of Geophysical Research-Atmospheres, 115,* D02114.

Sorooshian, S., Hsu, K., Gao, X., Gupta, H., Imam, B., & Braithwaite, D. (2000). Evolution of the PERSIANN system satellite-based estimates of tropical rainfall. *Bulletin of the American Meteorological Society, 81*(9), 2035–2046.

Tian, Y., Peters-Lidard, C. D., Eylander, J. B., Joyce, R. J., Huffman, G. J., Adler, R. F., Hsu, K., Turk, F. J., Garcia, M., & Zeng, J. (2009). Component analysis of errors in satellite-based precipitation estimates, *Journal of Geophysical Research, 114,* D24101. doi:10.1029/2009JD011949.

Wilks, D. (2006). *Statistical methods in the atmospheric sciences*. (2nd ed., p. 627) Burlington: Academic

Yilmaz, K., Hogue, T., Hsu, K., Sorooshian, S., Gupta, H., & Wagener, T. (2005). Intercomparison of rain gauge, radar and satellite-based precipitation estimates with emphasis on hydrologic forecasting. *Journal of Hydrometeorology, 6*(4), 497–517.

Zhou, C., (2008). A two-step estimator of the extreme value index. *Extremes, 11*(3), 281–302.

Chapter 3
Satellite Observations

3.1 MODIS

The Moderate resolution Imaging Spectroradiometer (MODIS) instrument onboard NASA's Earth Observing System (EOS) Aqua and Terra platforms with 36 spectral bands provides valuable information about atmosphere, land and oceans (Ackerman et al 1998). The low earth orbiting satellite at the altitude of 705 Km also gives us important insight into cloud micro-physical characteristics ranging in wavelength from 0.4 to 14.4 μm.

The Terra satellite orbits the Earth in a descending orbit passing the equator in the morning (at 10:30 am local time), while the Aqua follows an ascending orbit, passing the equator at 1:30 pm local time. The spatial resolution of the MODIS data is 250 m for visible channels (channels 1 and 2, 0.6–0.9 μm), 500 m for channels 3–7 (0.4–2.1 μm), and 1000 m for channels 8–36 (0.4–14.4 μm). MODIS has a swath width of 2330 Km and can span the entire surface of the Earth every 1 to 2 days. Terra launched in December 1999 and Aqua joined the EOS PM-1 in May 2002.

Figure 3.1 presents the bandwidth and primary usage of MODIS multi-spectral channel data (source: http://modis.gsfc.nasa.gov). Visible data are in the range of 400–700 nm and MODIS channels 1 and 2 are in this range. Visible channel images show the reflected solar radiation from the earth and atmosphere during daylight. Thick clouds, such as deep convective clouds, as well as ice and fresh snow on the earth's surface appear brightly on visible images. Water bodies such as lakes and oceans appear dark due to their low albedo. Surface features over land will be darker than clouds and brighter than water, but it might be very difficult to distinguish between low warm clouds and surface. Visible images are not strong detectors of thin clouds, such as cirrus formations.

The reflective infrared region of the electromagnetic spectrum has the bandwidth of 0.7–3 μm. Figure 3.1 presents the visible, reflective IR and thermal IR regions of the electromagnetic spectrum. The black shades on the Figure show regions that most of the energy is absorbed by the atmosphere. The white regions on the spec-

N. Nasrollahi, *Improving Infrared-Based Precipitation Retrieval Algorithms Using Multi-Spectral Satellite Imagery,* Springer Theses, DOI 10.1007/978-3-319-12081-2_3

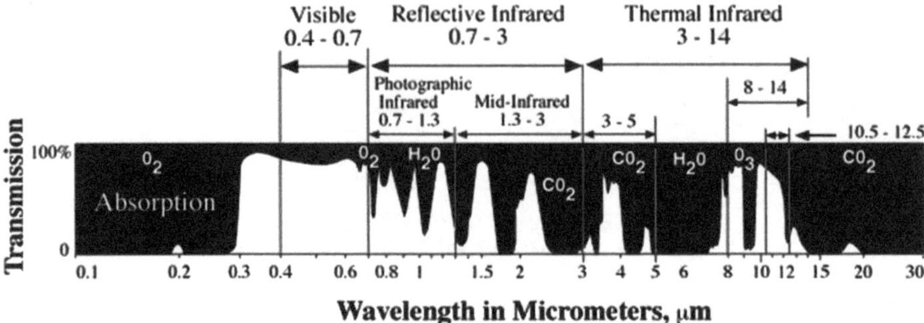

Fig. 3.1 Atmospheric windows in the electromagnetic spectrum. (Source: Jensen 2007)

trum are called the atmospheric window and the atmosphere passes part of the IR from the terrain to the satellite sensor (Jensen 2007).

A water vapor channel is an infrared channel in the range of 6.5–7.5 μm. Water vapor absorbs most of the radiation in this part of the spectrum. Most of the radiance received by the satellite in the water vapor channels comes from humidity that exists in the mid-upper troposphere. In these channels, surface features cannot be detected and only high clouds are recognizable. Water vapor channels are relatively noise-free that can show the movement of moisture in the atmosphere. The water vapor absorption regions are marked with H_2O in Fig. 3.1. The MODIS channels 27 and 28 are water vapor channels.

The wavelengths from 10.5 to 12.5 μm are thermal infrared regions that most part of the emitted energy from the terrain will be passed to the sensor with very limited absorption. Channels 31 and 32 of MODIS are sensitive to this range of the spectrum. Channel 31 centers at 11.03 μm and channel 32 centers at 12.02 μm.

In this study a set of 7 MODIS channels are used, one in the range of visible (channel 1), 2 water vapor channels (channels 27 and 28) and the rest are thermal infrared channels (channels 29, 30, 31 and 32). The details of microphysical properties of cloud and electromagnetic sensitivities are explained in Sect. 4.1 (Table 3.1).

For this study, the MODIS level 1B calibrated radiance data were used. The original radiance data from MODIS have the units watts per square meter per steradian per micrometer (Watts/m^2/micrometer/steradian). Radiance data are converted to Brightness Temperature (BT) in Kelvin using the following equation (Cohen and Taylor 1993):

$$
\begin{aligned}
&c_1 = 2.0 * h * c^2; \\
&c_2 = h * c / k; \\
&ws = 1.0^{-6} * \text{wavelength}; \\
&BT = c^2 ./ \left(ws * \log\left(c1 ./ \left(1.0^6 * \text{radiance} * ws^5 \right) + 1.0 \right) \right);
\end{aligned}
\tag{3.1}
$$

Table 3.1 Information about 36 spectral channels of MODIS instrument

Primary use	Band number	Central wavelength [nm]	Bandwidth[nm]	Spatial resolution [m]
Land/cloud/ aerosols/ boundaries	1	645	620–670	250
	2	858.5	841–876	250
Land/cloud/ aerosols properties	3	469	459–479	500
	4	555	545–565	500
	5	1240	1230–1250	500
	6	1640	1628–1652	500
	7	2130	2105–2155	500
Ocean color/ phytoplankton/ biogeochemistry	8	421.5	405–420	1000
	9	443	438–448	1000
	10	488	483–493	1000
	11	531	526–536	1000
	12	551	546–556	1000
	13	667	662–672	1000
	14	678	673–683	1000
	15	748	743–753	1000
	16	869.5	862–877	1000
Atmospheric water vapor	17	905	890–920	1000
	18	936	931–941	1000
	19	940	915–965	1000
Surface/cloud temperature	20	3750	3660–3840	1000
	21	3959	3929–3989	1000
	22	3959	3929–3989	1000
	23	4050	4020–4080	1000
Atmospheric temperature	24	4465.5	4433–4498	1000
	25	4515.5	4482–4549	1000
Cirrus clouds/ water vapor	26	1375	1360–1390	1000
	27	6715	6535–6895	1000
	28	7325	7175–7475	1000
Cloud properties	29	8550	8400–8700	1000
Ozone	30	9730	9580–9880	1000
Surface cloud temperature	31	11030	10780–11280	1000
	32	12020	11770–12270	1000
Cloud top altitude	33	13335	13185–13485	1000
	34	13635	13485–13785	1000
	35	13935	13785–14085	1000
	36	14235	14085–14385	1000

Where:

h = 6.6260755d−34; Planck constant (Joule second)
c = 2.9979246d + 8; Speed of light in vacuum (meters/second)
k = 1.380658d−23; Boltzmann constant (Joules/Kelvin)

3.2 CloudSat

CloudSat (a NASA Earth Sciences Systems Pathfinder (ESSP) mission) is designed to measure the vertical structure of clouds from space and provides the first direct observation of cloud vertical structure (Weisz et al. 2007). CloudSat is incorporated into the EOS satellites, which fly in a sun-synchronous orbit at a 705 Km altitude. The CloudSat satellite consists of a 94 GHz Cloud Profiling Radar (CPR) and provides a rich source of information about cloud properties. CloudSat data are available at resolution of 1.1 Km along track by 1.3 Km across track. All CloudSat data products are available to download from the CloudSat Data Processing Center (http://cloudsat.cira.colostate.edu). MODIS and CloudSat onboard Aqua are both part of the afternoon constellation of satellites, called the A-Train (Stephens et al. 2002). The A-Train formation (Fig. 3.2) currently consists of a set of 4 satellites, starting with Aqua and followed by CloudSat and CALIPSO, with Aura as the last satellite. The carbon-tracking Orbiting Carbon Observatory 2 (OCO-2) satellite will be launched in 2014 to provide space-based global measurements of atmospheric carbon dioxide (CO_2). The PARASOL (Polarization & Anisotropy of Reflectances for Atmospheric Sciences coupled with Observations from a Lidar) moved out of the A-Train in December 2009. The CloudSat radar flies in-formation with Aqua, with an average of 60 s delay between them, providing almost simultaneous observations.

There are several CloudSat products available from the CloudSat science team. Among them, the cloud scenario classification (2B-CLDCLASS) and Precipitation Column Algorithm Product (2C-PRECIP-COLUMN) are of interest to this research.

Different cloud types have different microphysical properties, frequency and dynamic forcing. Climate change can alter the frequency and properties of clouds, resulting in changes in precipitation occurrence and intensity.

Using space based observation of radar reflectivity from CPR and lidar observations available from CALIPSO, as well as MODIS radiances, Sassen and Wang (2008) developed the cloud classification algorithm. The CPR and lidar data are useful in identifying the vertical and horizontal extend of clouds, cloud temperature and the presence of precipitation (Wang and Sassen 2007).

CloudSat cloud-type classification product is able to identify clear sky, as well as 7 different classes of clouds: cumulus (Cu), stratocumulus (Sc), altocumulus (Ac), altostratus (As), nimbostratus (Ns), high cloud (cirrus or cirrostratus) and deep convective cloud. In the latest version of the CloudSat CLD_CLASS dataset, St and

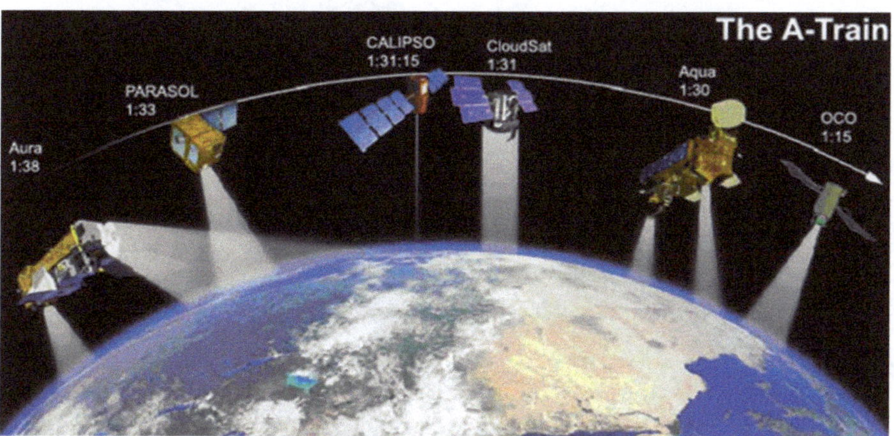

Fig. 3.2 The A-Train constellation. (Source: http://cloudsat.atmos.colostate.edu/education/satellites)

Sc clouds are combined in one group. The class of high cloud in the cloud scenario classification includes cirrus, cirrocumulus, and cirrostratus. Cirrus is high-level cloud that mainly consists of single ice particles. These white or light gray color clouds appear in elevations higher than 5 Km and their thickness changes between 100 to 8000 m. Cirrocumulus clouds are high level convective clouds and cirrostratus is extensive cirrus in high altitudes. High-level clouds have very cold cloud tops and clearly appear in the IR images. Cirrus clouds are very thin or semi-transparent and might not be distinguishable in the visible imagery. Cirrosratus clouds are very large in horizontal direction and have homogeneous texture. High clouds are one of the non-precipitating cloud groups.

Middle-level clouds, such as altostratus and altocumulus, can be distinguished in IR images because of their cold tops. They can be homogeneous (e.g. As) and inhomogeneous (e.g. Ac). Winter time detection of middle-level clouds might be challenging in high or mid-latitude regions and also in high mountainous regions.

Low-clouds are typically present in the elevations lower than 3 Km, such as cumulus, stratus and stratocumulus. Cumulus clouds are puffy shaped clouds that are vertically expanded. They appear lower in the atmosphere (lower that 2 Km) and have flat bases. Cumulus clouds usually produce very light or zero precipitation, but they can grow into cumulonimbus clouds that are precipitating cloud types. Stratus clouds in contrast have homogeneous texture and their horizontal extension is larger compared to cumulus clouds. Stratocumulus is another class of low clouds that are shallow, inhomogeneous and large horizontal dimension. In general, remote sensing detection of low clouds is challenging due to their warm cloud tops that appear close to surface radiation temperature in the IR channels.

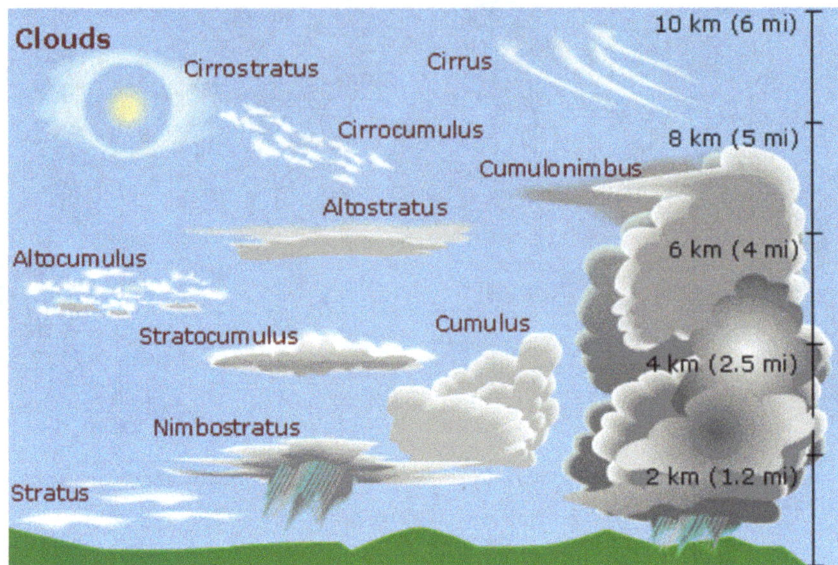

Fig. 3.3 Common types of clouds. (Source: http://airlineworld.files.wordpress.com/2008/07/cloud_types.gif)

The other group of clouds is deep clouds, such as nimbostratus and deep convective clouds (e.g. cumulonimbus). These two cloud types can extend from near the surface to the higher troposphere; however, their main difference is in their precipitation intensity. DC clouds produce heavier precipitation compared to NS clouds and are formed after strong updrafts. Figure 3.3 represents common types of clouds and Table 3.2 provides characteristics of different cloud scenarios provided by the CloudSat science team.

Table 3.2 Different cloud types characteristics. (Source: Wang and Sassen (2007))

Cloud class	Cloud features	
High cloud	Base	> 7.0 km
	Rain	No
	Horiz. dim.	10^3 km
	Vert. dim.	Moderate
	LWP	= 0
As	Base	2.0 – 7.0 km
	Rain	None
	Horiz. dim.	10^3 km, homogeneous
	Vert. dim.	Moderate
	LWP	0, dominated by ice
Ac	Base	2.0 – 7.0 km
	Rain	Virga possible
	Horiz. dim.	10^3 km, inhomogeneous
	Vert. dim.	Shallow or moderate
	LWP	> 0
St	Base	0 – 2.0 km
	Rain	None or slight
	Horiz. dim.	10^2 km, homogeneous
	Vert. dim.	Shallow
	LWP	> 0
Sc	Base	0 – 2.0 km
	Rain	Drizzle or snow possible
	Horiz. dim.	10^3 km, inhomogeneous
	Vert. dim.	Shallow
	LWP	> 0
Cu	Base	0 – 3.0 km
	Rain	Drizzle or snow possible
	Horiz. dim.	1 km, isolated
	Vert. dim.	Shallow or moderate
	LWP	> 0
Ns	Base	0 – 4.0 km
	Rain	Prolonged rain or snow
	Horiz. dim.	10^3 km
	Vert. dim.	Thick
	LWP	> 0
Deep convective	Base	0 – 3.0 km
	Rain	Intense rain or hail
	Horiz. dim.	10 km
	Vert. dim.	Thick
	LWP	> 0

References

Ackerman, S. A., Strabala, K. I., Menzel, W. P., Frey, R. A., Moeller, C. C., & Gumley, L. E. (1998). Discriminating clear sky from clouds with MODIS. *Journal of Geophysical Research, 103*(D24), 32141–32157. doi:10.1029/1998JD200032.

Cohen, E. R., & Taylor, B. N..(1993). The fundamental physical constants. *Physics Today, 45,* 9.

Jensen, R., Mausel, P., Dias, N., Gonser, R., Yang, C., Everitt, J., & Fletcher, R. (2007). Spectral analysis of coastal vegetation and land cover using AISA+ hyperspectral data. Geocarto International, *22*(1), 17–28.

Sassen, K., & Wang, Z. (2008). Classifying clouds around the globe with the CloudSat radar: 1 year of results. *Geophysical Research Letters, 35,* L04805. doi:10.1029/2007GL032591.

Stephens, G., et al. (2002). The CloudSat mission and the a-train: A new dimension of space- based observations of clouds and precipitation. *Bulletin of the American Meteorological Society, 83,* 1771–1790.

Wang, Z. K., & Sassen, K. (2007). *Level 2 cloud scenario classification product process description and interface control document.* Colorado State University: Cooperative Institute for Research in the Atmosphere.

Weisz, E., Li, J., Menzel, W. P. Heidinger, A., Kahn, B. & Liu, C.-Y. (2007). Comparisons of airs, modis, cloudsat and calipso cloud top height retrievals. *Geophysical Research Letters, 34,* 1–5.

Chapter 4
Reducing False Rain in Satellite Precipitation Products Using Cloudsat Cloud Classification Maps and Modis Multi-spectral Images

Because clouds play important roles in producing precipitation and in Earth's radiative balance, they are a key element in studies of weather and climate, water and energy cycles, and hydrologic analysis. Low clouds have an important effect on cooling the Earth, as they reflect sunlight back to space. High, thin clouds have the opposite effect, allowing incoming sunshine to pass through but trapping heat that is trying to escape from earth. Improving our understanding of cloud structures is the main step in global climate studies and precipitation algorithm development.

One of the unique observations available from CloudSat is its vertical cloud structure. CloudSat cloud classification data set is used in this study to classify non-precipitating clouds and therefore delineate the no-rain regions. By delineating no-rain areas, the false rain estimations from satellite precipitation algorithm will be removed. After explaining the role of multi-spectral data in precipitation algorithms, the classification methodology is discussed in detail.

4.1 The Role of Multi-spectral Data in Satellite Precipitation Algorithms

Many satellite-derived precipitation products take advantage of multiple remote sensing devices. For example, to overcome the temporal limitations of PMW estimates, NOAA CPC Morphing Technique (CMORPH) uses atmospheric motion vectors derived from GEO's IR data to propagate high quality PMW precipitation estimates when updated PMW data are unavailable (Joyce et al. 2004). TRMM Multi-satellite Precipitation Analysis (TMPA) products are combined precipitation products that use GEO's IR information to fill the gaps between PMW estimates (Huffman et al. 2007). Other precipitation products use PMW adjusted IR data, such as the PMW calibrated IR algorithm (PMIR; Kidd et al. 2003), the Precipitation Estimation from Remotely Sensed Information using Artificial Neural Networks (PERSIANN) algorithm (Hsu et al. 1997; Sorooshian et al. 2000), and the Self-Calibrating Multivariate Precipitation Retrieval algorithm (SCaMPR; Kuligowski 2002). In addition, the Naval Research Laboratory (NRL) blended-satellite

N. Nasrollahi, *Improving Infrared-Based Precipitation Retrieval Algorithms Using Multi-Spectral Satellite Imagery,* Springer Theses, DOI 10.1007/978-3-319-12081-2_4

precipitation technique uses a combination of MODIS/AMSR-E sensors to detect cirrus clouds and reduce false rain estimations in the algorithm (Turk and Miller 2005). More recently, Rain Estimation using the Forward-Adjusted advection of Microwave Estimates (REFAME) algorithm (Behrangi et al. 2010) uses IR images to advect microwave-derived rain rates along the cloud motion tracks. This algorithm takes advantage of a local cloud classification method to adjust the rain rates. More sophisticated approaches such as The Lagrangian Model (LMODEL) algorithm, combine information from microwave calibrated data and morphing techniques using a conceptual modeling framework (Bellerby et al. 2009; Hsu et al. 2009).

Several studies emphasize that more advanced methods are needed to improve the quality of satellite precipitation products, including reducing their FAR (Sorooshian et al. 2011; AghaKouchak et al. 2009). The utility of multi-spectral satellite data in capturing microphysical properties of clouds and improving precipitation estimation has been the subject of many investigations in recent years. For instance, Li et al. (2007) showed the effectiveness of MODIS channel 31 (11.03 μm) in identifying high clouds with very cold brightness temperatures. Strabala et al. (1994) show that for high ice clouds, a difference between 8.5 and 11 μm brightness temperatures (BTD[8.5-11]) is greater than BTD[11-12]. Furthermore, Wang et al. (2009) used the near-infrared (NIR) 2.19 μm band to retrieve cloud particle size and used the water vapor absorption channel 1.38 μm band to screen out upper-level ice clouds. Turk and Miller (2005) show that significantly positive BTD[3.7-11] provides information for identifying cirrus clouds at night.

BTD[11-12] is also useful in identifying ice clouds. Inoue (1987) showed that optically thin (τ in the range of 0.1 and 4) cirrus clouds have BTD[11-12] values greater than 2.5K. Furthermore, BTD[11-12] values less than or equal to 0K correspond to deep convective clouds with heavy precipitation (Kurino 1997). More recently, Setvak et al. (2003) showed that convective storms exhibit a significant increase in 3.7 μm cloud top reflectivity.

BTD[8.5-11] also has been shown to be effective in identifying high ice clouds. Since ice particles absorb much less radiation at 8.5 μm than 11 μm, high cirrus clouds are expected to have a BTD[8.5-11] greater than one (Roskovensky and Liou 2003). Thies et al. (2008) considered BTD[8.7-10.8] and BTD[10.8-12.1] to identify cloud phase.

Using multi-spectral data for rain/no-rain (R/NR) detection was also a focus of many studies. A combination of VIS and IR channels was initially used by Lovejoy and Mandelbrot (1985) and Austin (1987) to identify R/NR occurrences. Capacci and Conway (2005), Behrangi et al. (2010), and others have also found remarkable improvements in detecting rainy areas when using multi-spectral data. Lensky and Rosenfeld (2003) implemented the difference between a thermal IR channel and a mid-IR channel, BTD[3.7-11], into a night-rain delineation algorithm. Kwon et al. 2009 shows improvements in detecting deep convective cloud heights by using Ozone channel 9.7μm (MODIS channel 30).

4.2 Satellite Data

The proposed method will benefit from more reliable and detailed information on cloud classes, obtained from CloudSat satellite, to differentiate precipitating and non-precipitating cloud types. Identifying high cold clouds helps to screen out non-precipitating clouds, and therefore reduce FAR in current precipitation algorithms. Among various types of clouds, cirrus (high) and altostratus clouds are non-precipitating clouds that are the interest of this study.

Not all 36 channels of MODIS are beneficial in cloud and precipitation studies. As explained in Sect. 4.1 among different multi-spectral channels there are a few of them that are useful in precipitation and cloud detection algorithms. Those channels are in the range of water vapor and infrared. In this study, a set of six WV and IR channels of MODIS (6.75, 7.325, 8.55, 9.7, 11.03 and 12.02 μm) were selected as input to the ANN model. The availability of these channels during the day and night makes it possible to have a consistent rain/no-rain (R/NR) detection algorithm for day and night retrieval.

4.3 Methodology

To give an example of how different datasets are used in this study, Fig. 4.1 is presented.

Figure 4.1a demonstrates the CloudSat overpass through a precipitation event (Stage IV data) over South Carolina and neighboring states on August 13th, 2008 (05:45 UTC). The black line in Fig. 4.1a represents the track of the CloudSat radar, while the second panel in the figure shows the vertical profile of the clouds with different cloud types obtained from the 2B-CLDCLASS product. The X axis shows the pixel number along the track of CloudSat (each pixel is approximately 1.1 Km). The Y axis shows the cloud height in Km. As demonstrated in this figure, high clouds (blue color) have cloud tops higher than 12 Km and deep convective clouds (shown in brown) are very thick and have high cloud tops. Furthermore, the figure displays PERSIANN (panel c) precipitation estimates and radar observations (panel d) corresponding to the CloudSat track. The PERSIANN and Stage IV data are 3 hourly data and CloudSat has instantaneous observation. In this example the data are chosen in a way to have the minimum time difference between CloudSat observation at 5:45 and PERSIANN and Stage IV most probably at 5:45 and 6:00 respectively. Considering panels c and d, one can see that the maximum amount of precipitation estimated by PERSIANN coincides with the high cirrus anvil, which has the lowest brightness temperature. However, ground-based data indicates that the peak of the storm is in the center of the deep convective tower (about 15 mm/hr), which makes more physical sense. Radar data estimates zero precipitation in the presence of high clouds. Panels (e) and (f) in Fig. 4.1 display cloud brightness temperature converted from MODIS radiance data, which are informative in terms of different cloud types. Panel (e) shows that the lowest value of brightness temperature at 11 μm appears at the location of high clouds, and coincides with high precipitation estimates from

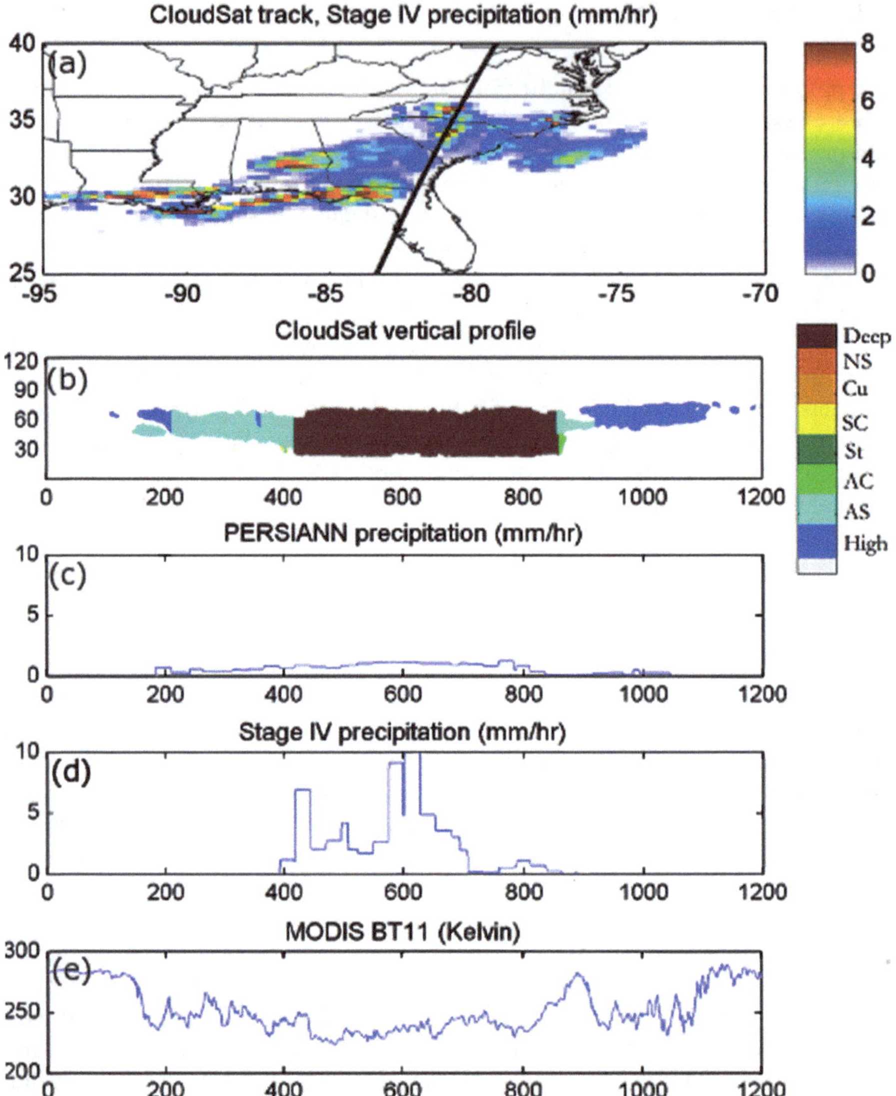

Fig. 4.1 Part a. An example of the CloudSat cloud classification map and MODIS brightness temperature data on August 13th, 2008 (05:45 UTC). **a** Track of CloudSat passing through a storm measured by Stage IV precipitation data. **b** CloudSat vertical cloud profile. **c** PERSIANN precipitation data (mm/hr). **d** Stage IV precipitation data (mm/hr). **e** MODIS brightness temperature at 11 μm (*Kelvin*). Part b. **f** MODIS BTD[8.5–11] (*Kelvin*). **g** Radar reflectivity (*dBZ*)

the PERSIANN product. As discussed earlier, the brightness temperature difference between channels 31 and 29 of MODIS (BTD[8.5-11]) is a strong positive value (greater than 2 K) for high ice clouds. Panel (g) in the figure presents the radar reflectivity observations by CloudSat showing the vertical structure of the convective zone. MODIS BTD[8.5–11] is almost zero in the presence of deep convective cloud

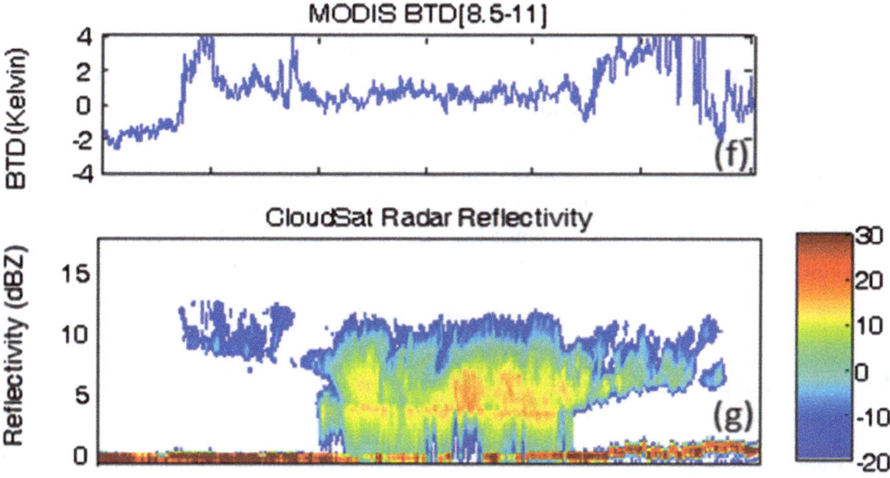

Fig. 4.1 (continued)

as shown in panels (f) and (g). The distinction between optically thin clouds (i.e. cirrus) and optically deep clouds (i.e. convective clouds) from multi-spectral channels helps to improve the IR only algorithms. Underestimation of PERSIANN algorithm in the presence of deep-convective clouds is one of the limitations of IR-based algorithms.

Figure 4.2 presents how the false rain identification algorithm works. In this method, the cloud classes obtained from CloudSat are assigned to textural and spectral features of clouds observed by MODIS, whenever CloudSat retrieval is

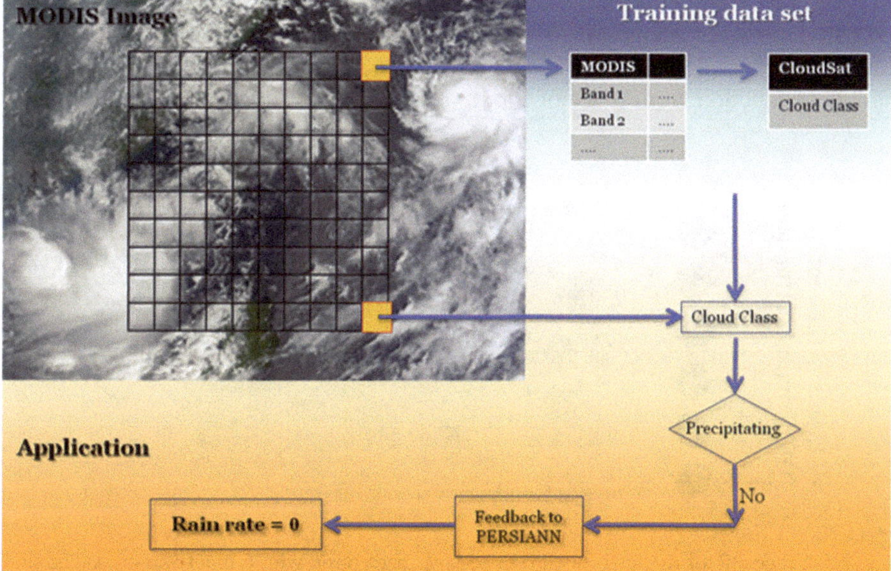

Fig. 4.2 False rain identification algorithm using cloud classification data)

available. A training data set is created from CloudSat and MODIS data over the continental United States. At each CloudSat track, pixels with single layer cloud are identified and the cloud class with cloud multi-spectral information from MODIS is stored in the training data matrix. The target value in the target vector is in the binary format. If the cloud class is a non-precipitating cloud type (i.e. high and altostratus), the target value is 1, and if the pixel is associated with other types of precipitating clouds, the target value is 0. The training database is then used as a reference to find the best cloud class for the times that CloudSat data is not available. In the following section, the details of the classification method are explained.

4.4 Classification

Multi-spectral image classification is an important technique in the application of remote sensing and geo-sciences. Statistical classification is a multivariate analysis that takes advantage of simultaneous observations coming from images on different spectral bands. Analyzing a set of input variables for a set of known classes (i.e. labels), a statistical connection is created between the input features and the target response (i.e. training data set). Among different classification techniques, Artificial Neural Networks (ANNs) have been shown to be an effective tool in classifying complicated systems (e.g. Hsu et al. 1997; Capacci and Conway 2005; Behrangi et al. 2009; Hong et al. 2004; Farahmand and AghaKouchak 2013; Bellerby et al. 2000; Tapiador et al. 2004).

ANNs are pattern recognition tools usually used to model complex relationships between a set of inputs and corresponding outputs (Bishop 1996). These models are composed of interconnecting artificial neurons, and are employed to find statistical correlations between multi-spectral information on cloud tops and binary target value (see Fig. 4.3 for ANNs' model structure). In other words, ANN is simply a nonlinear function from a set of input variables (x) to a set of output variables

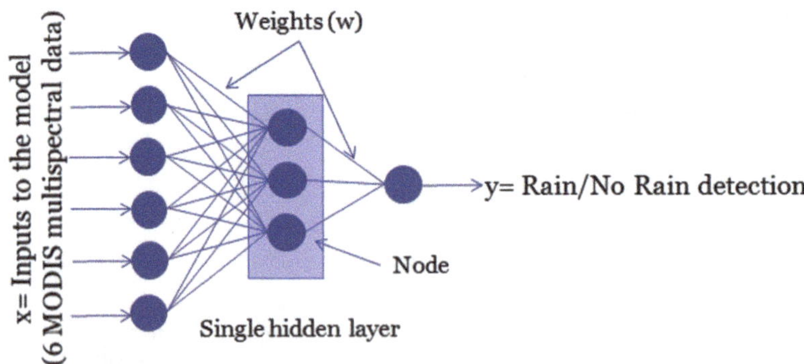

Fig. 4.3 Schematic of the feed-forward three-layer perceptron with 6 input variables. The final output layer provides the rain/no rain detection)

(target values, y) with a vector of adjusted parameters. (ω) ANNs are capable of mapping multivariate functions and of extracting underlying rules from noisy data. In addition, they are well suited to problems of estimation and prediction in hydrometeorology and remote sensing. They are popular for estimating and forecasting precipitation (Hsu et al. 1997, 1999; Sorooshian et al. 2000; Hong et al. 2004; Behrangi et al. 2010), and for some other remote sensing image classification and applications (Benediktsson et al. 1990; Hara et al. 1995; Ji 2000; Aitkenhead and Dyer 2007). ANNs can approximate any continuous input-output function, and its derivatives, to arbitrary accuracy (Hornik et al. 1990; Gallant and White 1992).

In this study, a feed-forward back-propagation model with a single hidden layer and a sigmoidal activation function was created. The ANN model calculates the errors between the calculated output and given output data, and by adjusting the weights, minimizes the error. The general equation for ANNs is in the form of a linear combination of fixed nonlinear basis functions, $\varphi_j(x)$, with the weights ω_j and is in the form of:

$$y(x,w) = f\left(\sum_{j=1}^{M} \omega_j \phi_j(x)\right)$$

Each basis function, $\varphi_j(x)$, itself is a nonlinear function of a linear combination of the inputs (i.e. MODIS data), where the coefficients in the linear combination are parameters to be adjusted during model training.

In the general ANN equation, f is the activation function. In this study a sigmoidal activation function was associated with all the neurons in the model, and is in the form of:

$$f = \frac{1}{1+e^{-a}}$$

The target values in the ANN model are a binary vector of no-rain clouds (1) or possible raining clouds (0). The ANN computes the value of the output based on the series of inputs entered into the model. If the output is equal or greater than 0.5, it assumed to be a no-rain scenario, while values less than 0.5 are possible rain pixels.

4.5 Training Data Set

Six MODIS infrared and water vapor channels (6.75, 7.325, 8.55, 9.7, 11.03, 12.02 μm wavelength) are set as input variables to the ANN model to recognize different pattern of rain and no-rain clouds. The training data set was created by obtaining information about MODIS multi-spectral data and CloudSat cloud class data, whenever CloudSat retrieval was available. Training data consists of 150,000 cloudy pixels in summer 2008. A one layer feed forward, back propagation neural network model was employed to identify no-rain clouds. The target to the ANN model is a binary matrix, having one when there is a non-precipitating cloud and zero otherwise. For the times that CloudSat is not available, the trained model is used to find the non-precipitating cloud coverage in each MODIS image.

4.6 Application of the Model on Precipitation Events

Figure 4.4a shows precipitation estimation by Stage IV precipitation for Aug. 13, 2008 (0545 UTC). Panel (b) on the figure represents corresponding PERSIANN estimation (mm/hr). Figure 4.4c demonstrates the false alarm precipitation that can be removed using the proposed method. Data in panel (c) are from 2 consequent MODIS granules with about 5 min delays from each other (region between solid black lines). Pixels with false precipitation are shown with red and blue. The blue color highlights the no-rain pixels that are falsely identified as rain pixels in the PERSIANN dataset; however, the algorithm identifies them as associated with non-precipitating cloud (cirrus). The red color shows the false rain pixels not identified with this algorithm. From the total number of pixels with false precipitation, more than 55 % of pixels are identified with the current method. Figure 4.5 shows the results for the event on Jul. 22, 2008 (0805 UTC). The false rain pixels identification rate is 43 % in this event.

4.7 Results and Discussions

High false rain in IR-based satellite precipitation algorithms is one of their well-known shortcomings. Using the multi-spectral images of MODIS satellite and unique capability of CloudSat in observing vertical structure of clouds, a new level of information can be added to the current precipitation algorithms. By identifying non-precipitating clouds, the regions of no-rain can be identified and therefore the false rain area in satellite precipitation algorithms will be determined.

A training algorithm using ANN method and unique observations of CloudSat and MODIS is created to identifying regions associated with non-precipitating clouds. The two examples presented in this research clearly show the improvements gained by adding new sources of information to the current rain algorithm.

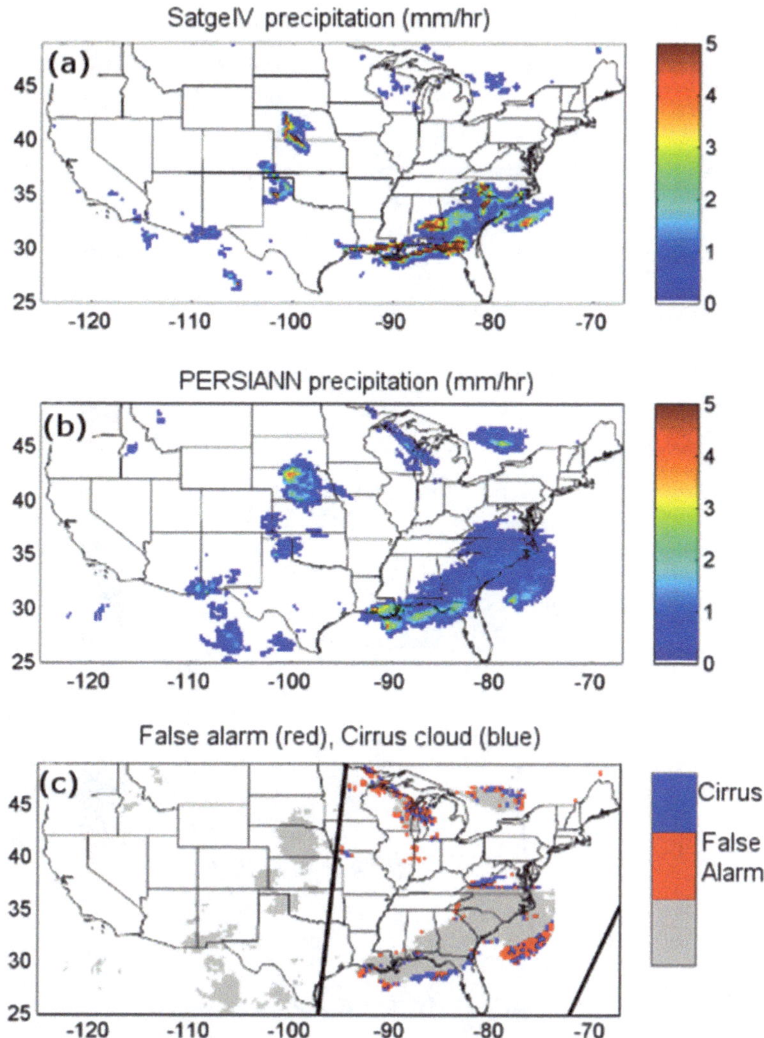

Fig. 4.4 False alarm detection using MODIS 6 spectral bands and CloudSat CLD-CLASS August 13, 2008 (0545 UTC))

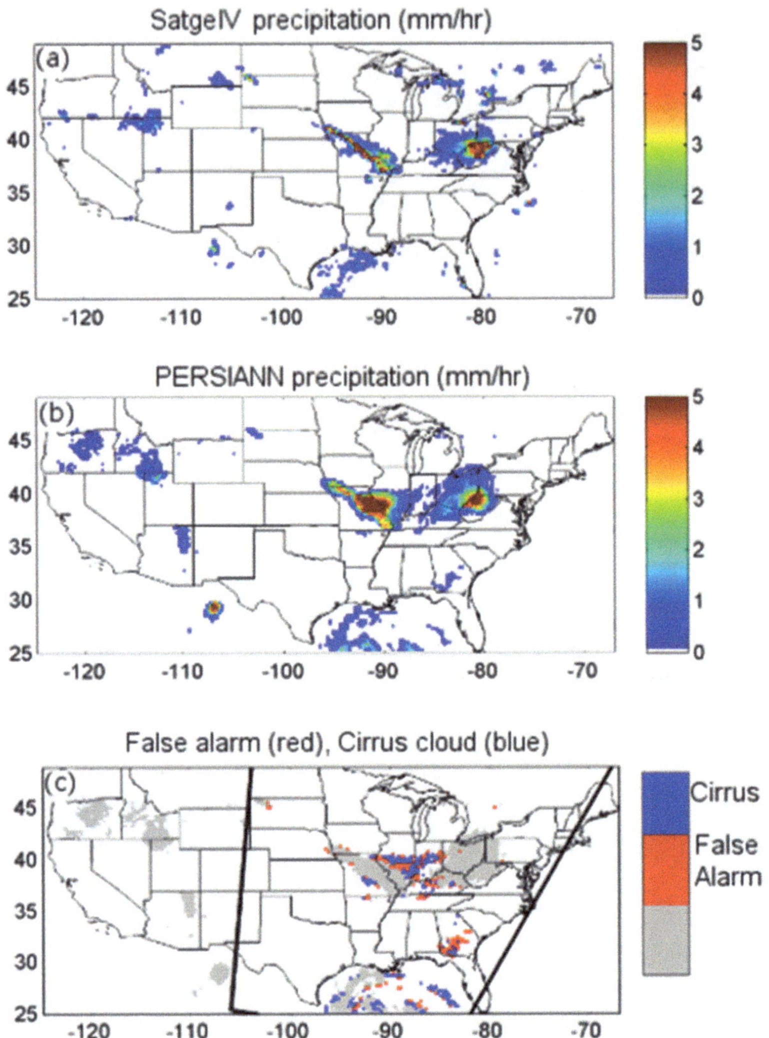

Fig. 4.5 False alarm detection using MODIS 6 spectral bands and CloudSat CLD-CLASS, July 22, 2008 (0805 UTC))

References

AghaKouchak, A, Nasrollahi, N, Habib, E, (2009). Accounting for Uncertainties of the TRMM Satellite Estimates, Remote Sensing, 1(3), 606–619, doi:10.3390/rs1030606.

Aitkenhead, M.J., & Dyer, R. (2007). Improving land-cover classification using recognition threshold neural networks. *Photogrammetric Engineering and Remote Sensing, 73*(4), 413–421.

Austin, P. (1987). Relation between measured radar reflectivity and surface rainfall. *Monthly Weather Review, 115,* 1053–1070.

Behrangi, A., Hsu, K., Imam, B., Sorooshian, S., & Kuligowski, R. 2009: Evaluating the utility of multi-spectral information in delineating the areal extent of precipitation. *Journal of Hydrometeorology, 10*(3), 684–700.

Behrangi, A., Hsu, K., Imam, B., & Sorooshian, S. (2010a). Daytime precipitation estimation using bispectral cloud classification system. *Journal of Applied Meteorology, 49,* 1015–1031.

Behrangi, A., Imam, B., Hsu, K., Sorooshian, S., Bellerby, T. H., & Huffman, G. J. (2010b). REFAME: Rain estimation using forward-adjusted advection of microwave estimates. *Journal of Hydrometeorology, 11,* 1305–1321.

Bellerby, T., Todd, M., Kniveton, D., & Kidd, C. (2000) Rainfall estimation from a combination of TRMM precipitation radar and goes multi-spectral satellite imagery through the use of an artificial neural network. *Journal of Applied Meteorology, 39,* 21152128.

Bellerby, T., Hsu, K., & Sorooshian, S. (2009) Lmodel: A satellite precipitation methodology using cloud development modeling. Part I: Algorithm construction and calibration. *Journal of Hydrometeorology, 10,* 1081–1095.

Benediktsson, J. A., Swain, P.H., Ersoy, O. K. (1990). Neural network approaches versus statistical-methods in classification of multisource remote-sensing data. *IEEE Transactions on Geoscience and Remote Sensing, 28*(4), 540–552.

Bishop, C. (1996). *Neural networks for pattern recognition.* Oxford: Oxford University Press.

Capacci, D., & Conway, B. (2005). Delineation of precipitation areas from MODIS visible and infrared imagery with artificial neural networks. *Meteorological Applications, 12,* 291–305.

Farahmand, A., AghaKouchak, A. (2013). A Satellite-Based Global Landslide Model, Natural Hazards and Earth System Sciences, 13, 1259–1267, doi:10.5194/nhess-13-1259-2013.

Gallant, A. R., & White, H. (1992). On learning the derivatives of an unknown mapping with multilayer feed forward networks. *Neural Networks, 5,* 129–138.

Hara, Y., Atkins, R. G., Shin, R. T., Kong, J. A., Yueh, S. H., & Kwok, R. (1995). Application of Neural Networks for Sea Ice Classification in Polarimetric SAR Images. *IEEE Transactions on Geoscience and Remote Sensing, 33*(3), 740–748.

Hong, Y., Hsu, K., Gao, X., & Sorooshian, S. (2004). Precipitation estimation from remotely sensed imagery using artificial neural network-cloud classification system. *Journal of Applied Meteorology and Climatology, 43,* 18341853.

Hornik, K., Stinchcombe, M., & White, H. (1990). Universal approximation of an unknown mapping and its derivatives using multi-layer feed-forward networks. *Neural Networks, 3,* 551–560.

Hsu, K., Gao, X., Sorooshian, S., & Gupta, H. (1997). Precipitation estimation from remotely sensed information using artificial neural networks. *Journal of Applied Meteorology, 36,* 1176–1190.

Hsu, K., Gupta, H., Gao, X., Sorooshian, S. (1999). Estimation of physical variables from multichannel remotely sensed imagery using a neural network: Application to rainfall estimation. *Water Resources Research, 35*(5), 1605-1618.

Hsu, K., Bellerby, T., & Sorooshian, S. (2009). Lmodel: A satellite precipitation methodology using cloud development modeling. Part ii: Validation. *Journal of Hydrometeorology, 10,* 1096–1108.

Huffman, G., Adler, R., Bolvin, D., Gu, G., Nelkin, E., Bowman, K., Stocker, E., & Wolff, D. (2007). The TRMM multi-satellite precipitation analysis: Quasi-global, multiyear, combined-sensor precipitation estimates at fine scale. *Journal of Hydrometeorology, 8,* 3855.

Inoue, T. (1987). A cloud type classification with NOAA 7 split-window measurements. *Journal of Geophysical Research-Atmospheres, 92,* 3991 4000.

Ji, C.Y. (2000). Land-Use Classification of Remotely Sensed Data Using Kohonen Self-organizing Feature Map Neural Networks. *Photogrammetric Engineering and Remote Sensing, 66*(12), 1451-1460.

Joyce, R., Janowiak, J., Arkin, P., & Xie, P. (2004). CMORPH: A method that produces global precipitation estimates from passive microwave and infrared data at high spatial and temporal resolution. *Journal of Hydrometeorology, 5,* 487–503.

Kidd, C., Kniveton, D., Todd, M., & Bellerby, T. (2003). Satellite rainfall estimation using combined passive microwave and infrared algorithms. *Journal of Hydrometeorology, 4,* 1088–1104.

Kuligowski, R. (2002). A self-calibrating real-time goes rainfall algorithm for short-term rainfall estimates. *Journal of Hydrometeorology, 3,* 112–130.

Kurino, T. (1997). A satellite infrared technique for estimating deep/shallow precipitation. *Satellite Data Applications: Weather and Climate, 19,* 511–514.

Kwon, E. H., Sohn, B. J., Schmetz, J., & Watts, P. (2009). Use of ozone channel measurements for deep convective cloud height retrievals over the tropics, Fifth Annual Symposium on Future Operational Environmental Satellite Systems- NPOESS and GOES-R. http://ams.confex.com/ams/89annual/techprogram/paper_148113.htm. Accessed 10 Jan 2014.

Lensky, I., & Rosenfeld, D. (2003). A night-rain delineation algorithm for infrared satellite data based on microphysical considerations. *Journal of Applied Meteorology, 42*(9), 1218–1226.

Li, Z., Li, J., Menzel, W., Schmit, T., & Ackerman, S. (2007) Comparison between current and future environmental satellite imagers on cloud classification using MODIS. *Remote Sensing of Environment, 108,* 311–326.

Lovejoy, S., & Mandelbrot, B. (1985) Fractal properties of rain, and a fractal model. *Tellus Series A- Dynamic Meteorology and Oceanography, 37*(3), 209–232.

Roskovensky, J., & Liou, K. (2003). Detection of thin cirrus using a combination of 1.38- μm reflectance and window brightness temperature difference. *Journal of Geophysical Research, 108,* AAC4–1–16.

Setvak, M., Rabin, R., Doswell, C., & Levizzani, V. (2003). Satellite observations of convective storm tops in the 1. 6, 3.7 and 3.9 μm spectral bands. *Journal of Atmospheric Research, 67-8,* 607–627.

Sorooshian, S., Hsu, K., Gao, X., Gupta, H., Imam, B., & Braithwaite, D. (2000). Evolution of the PERSIANN system satellite-based estimates of tropical rainfall. *Bulletin of the American Meteorological Society, 81*(9), 2035–2046.

Sorooshian, S., et al. (2011) Advanced concepts on remote sensing of precipitation at multiple scales. *Bulletin of the American Meteorological Society*. doi: 10.1175/2011BAMS3158.1.

Strabala, K., Ackerman, W., & Menzel, W. P. (1994). Cloud properties inferred from 8-12-μm data. *Journal of Applied Meteorology, 33,* 212 229.

Tapiador, F., Kidd, C., Levizzani, V., & Marzano, F. (2004). A neural networks-based fusion technique to estimate half-hourly rainfall estimates at 0.1 degrees resolution from satellite passive microwave and infrared data. *Journal of Applied Meteorology, 43*(4), 576–594.

Thies, B., Nauss, T., & Bendix, J. (2008). Discriminating raining from non-raining cloud area at mid-latitudes using MSG SEVIRI daytime data. *Atmospheric Chemistry and Physics, 8,* 2341–2349.

Turk, F., & Miller, S. (2005). Toward improving estimates of remotely sensed precipitation with MODIS/AMSR-E blended data techniques. *IEEE Transactions on Geoscience and Remote Sensing, 43,* 1059–1069.

Wang, X., Liou, K. N., Ou, S. S. C., Mace, G. G., & Deng, M. (2009). Remote sensing of cirrus cloud vertical size profile using MODIS data. *Journal of Geophysical Research-Atmospheres, 114,* 1–16.

Chapter 5
Integration of CloudSat Precipitation Profile in Reduction of False Rain

This study develops a no-rain detection algorithm that takes advantage of CloudSat and MODIS observations to detect no-rain areas. The CloudSat surface precipitation occurrence data set is a reliable source to detect rain or no-rain based on CloudSat radar data. The backscatter of radar data due to presence of hydrometeors near the surface confirms the occurrence of rain. In this chapter, the CloudSat precipitation occurrence is used as a reliable source for rain detection. After explaining the methodology and data sources, the model training and results are presented.

5.1 Classification

The ANN classification method is used to train the algorithm. Details of the ANN method for classification are explained in Sect. 4.4. The same model structure is used for this chapter. The main difference is the target value in the training algorithm. An ANN model with 20 nodes is created and their weights for each of the six MODIS window and infrared channels are presented in Fig. 5.1. Higher weight values show stronger input on that specific node. Figure 5.1 shows that different channels have different weights for summer and winter seasons.

5.2 Satellite Observations

A set of 6 MODIS infrared and water vapor channels (6.75, 7.325, 8.55, 9.7, 11.03, 12.02 μm wavelength) are considered as the input variables to the ANN model. The CloudSat Level 2-C Precipitation Column algorithm (Haynes et al. 2011) provides information about the presence of surface precipitation. The determination of surface precipitation occurrence is based on the radar reflectivity data near the surface and the surface reflection characteristics. Higher radar reflectivity near the surface increases the probability of rain near the surface. The CloudSat Precip. flag data set,

N. Nasrollahi, *Improving Infrared-Based Precipitation Retrieval Algorithms Using Multi-Spectral Satellite Imagery*, Springer Theses, DOI 10.1007/978-3-319-12081-2_5

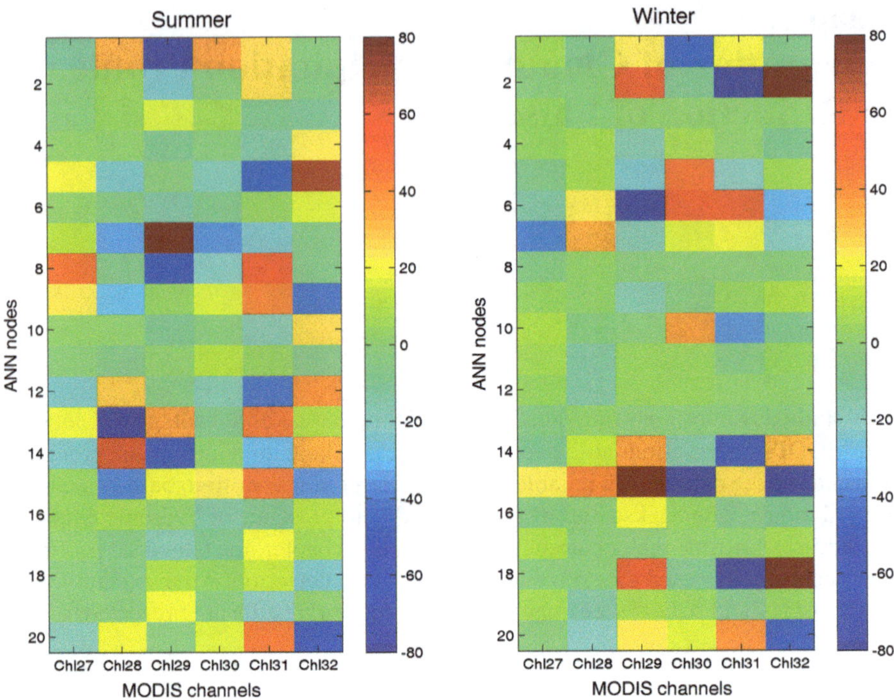

Fig. 5.1 The ANN model weights for summer (*left*) and winter (*right*) seasons

determines the surface rain occurrence based on the reflectivity values below 2 km altitude (Haynes et al. 2011).

The flag categorizes precipitation into 9 different groups: no precipitation, uncertain, rain possible, rain probable, rain certain, snow possible, snow certain, surface mixed precipitation, mixed precipitation possible and mixed precipitation certain. In this study, only instances of certain no-precipitation were considered as NR pixels.

5.3 Training Data Set

To have a better estimation of performance of the proposed technique, the analysis was done for summer and winter precipitation events. Separate training for summer and wintertime were considered to account for different climate conditions in different seasons and improve the accuracy of the model. As explained earlier, the spectral information from MODIS onboard Aqua and the corresponding CloudSat estimation of R/NR were considered in the training data sets. Data were randomly divided into two groups: training and testing. The summer training data included about 118,000 pixels observed on the summer of 2008, with 16,000 rainy pixels (dry to wet ratio of 7.3:1). Similarly, winter training data set with a dry to wet ratio of 2.8:1 embraced around 130,000 pixels in total from the winter of 2010.

5.4 Application of the Model on Precipitation Events

After training the algorithm using collocated MODIS and CloudSat pixels, the ANN model was used on MODIS multi-spectral images to identify the NR regions. At each MODIS pixel, the ANN model estimated if that pixel is a NR pixel, and the results were compared with CloudSat detections. The model performance was investigated over the continental United States for the summer and winter of 2007.

5.5 Results and Discussions

After training the model using the summer of 2008 and the winter of 2010 datasets, the model validation was performed on 2007 data. CloudSat radar data is considered as the truth to validate the R/NR classification model presented in this study. The 2007 summer results were evaluated over 70,000 CloudSat pixels and showed a 78 % accuracy in the detection rate of NR pixels. The 2007 winter data validation on 50,000 pixels showed a very high accuracy of 93 %. Figures 5.2 and 5.3 display the distribution of different cloud types for correct NR pixel classification, as well as the misclassified pixels for summer and the winter seasons, respectively.

Figure 5.2 shows high clouds and altostratus are two non-precipitation cloud types based on the CloudSat cloud classification algorithm. Most of the misclassified NR pixels are from the cloud types of altostratus (As) and altocumulus (Ac). Thirty nine percent of the pixels covered by altostratus clouds were misclassified in NR detection and the misclassification was around 34 % in the case of altocumulus clouds (5.1). The model's low performance in the case of middle level clouds, confirms the limitation of IR based algorithms in detecting warm clouds.

The distribution of different cloud classes in winter are demonstrated in Fig. 5.3. The first panel in the figure shows that most NR pixels are associated with high clouds and stratocumulus. The algorithm has the poorest performance in the case of middle level clouds, such as altostratus and altocumulus (see Table 5.1). The same result was observed in summer season classification. The misclassification rate is 14 % in the case of altostratus clouds and the error is less than 6 % in the remaining types of clouds. Model performance also depends on the number of pixels in the training dataset. The very poor model performance in the presence of deep convective clouds, with 43 % detection error in the winter season, is most likely due to insufficient number of pixels in the training dataset. In summer season, nimbostratus and deep convective clouds have the least occurrence. Most no-rain pixels that are associated with deep convective clouds and nimbostratus in summer are misclassified as rainy pixel in both summer and the winter seasons.

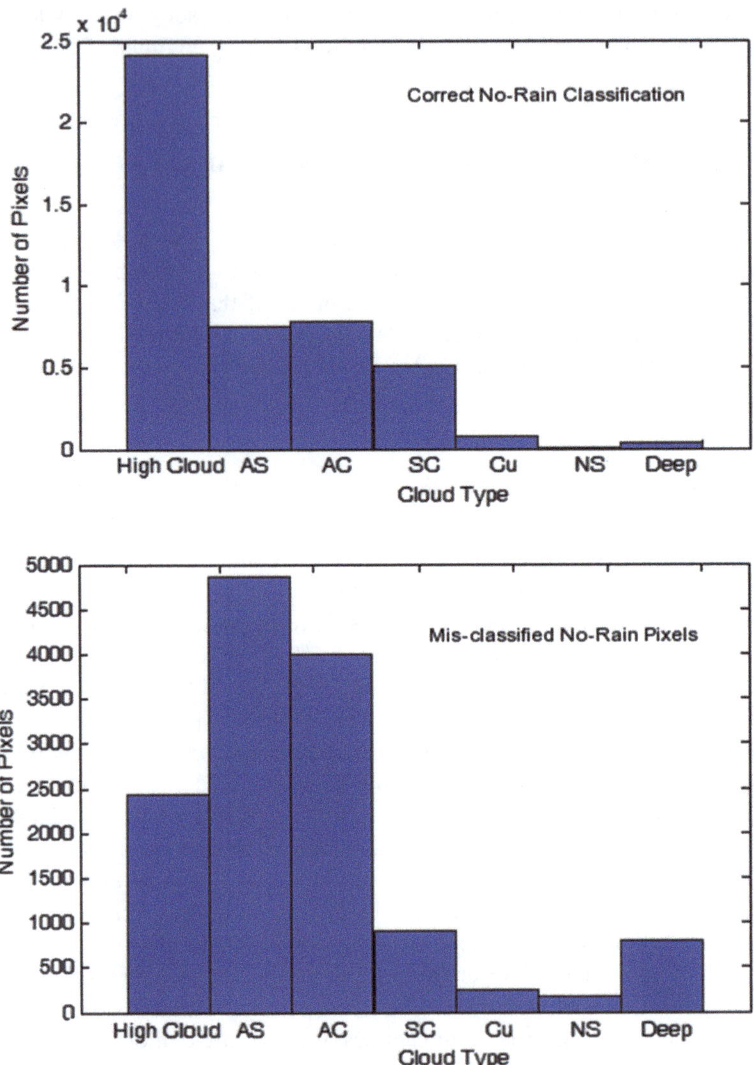

Fig. 5.2 Distribution of different cloud types in the case of correct NR detection (*top*) and misclassifications (*bottom*), for the summer of 2007

As discussed in the first section, the PERSIANN dataset shows higher false alarms in the winter season (Sorooshian et al. 2011). Applying the current algorithm, one can see an improvement of precipitation estimation in the winter season.

During summertime NR pixels associated with high clouds are identified with high accuracy. As discussed earlier, high clouds account for the majority of false rain estimations in satellite rainfall algorithms. Therefore, using the proposed methodology has a significant role in reducing false rain. We also acknowledge that the temporal differences between different datasets (i.e. MODIS and GOES observations) could affect the results.

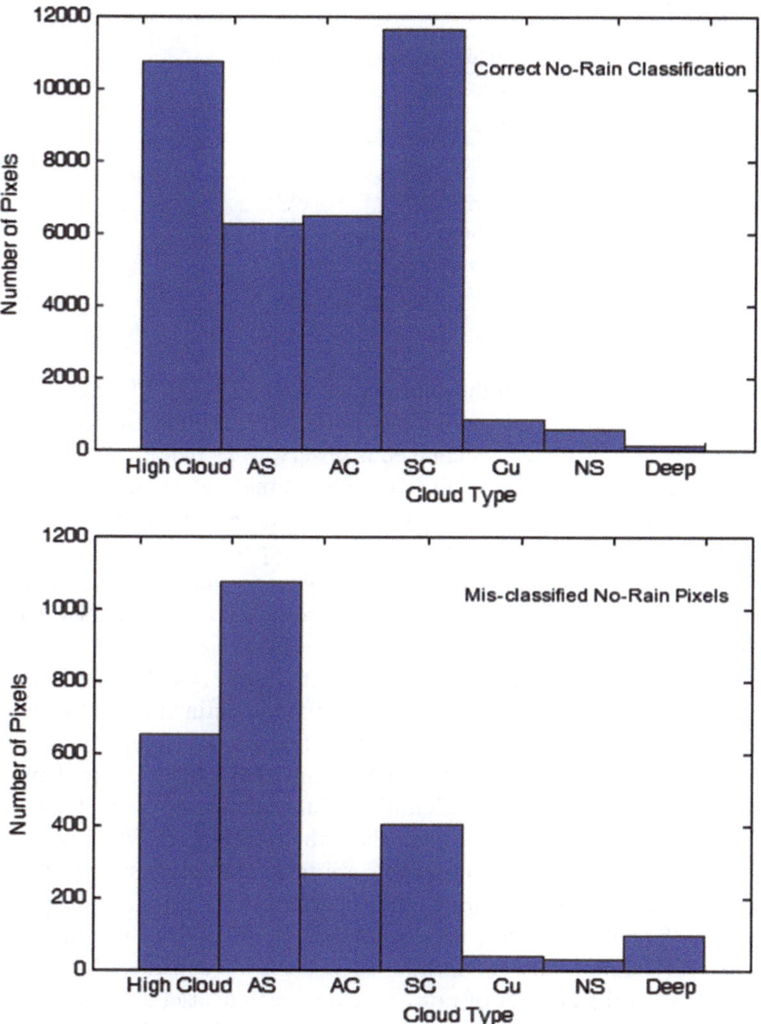

Fig. 5.3 Distribution of different cloud types in the case of correct NR detection (*top*) and misclassifications (*bottom*), for the winter of 2007

5.6 Case Study

Two case studies on summer and winter precipitation events are presented here to show the application of this technique to improving the quality of satellite rain estimation. The MODIS level 1B data set has a spatial resolution of 1 km in contrast to the 0.25° (~25 km) PERSIANN precipitation product. Therefore, the MODIS images were re-gridded to the 0.25° PERSIANN grids and then used as input to the ANN model.

Table 5.1 NR misclassification error percentage for summer and winter seasons

	Misclassification error percentage (%)	
Cloud type	Summer	Winter (%)
High cloud	9	6
Altostratus	39	14
Altocumulus	34	4
Stratocumulus	15	3
Cumulus	24	4
Nimbostratus	72	5
Deep convective	68	43

The temporal resolutions of the data sets are also different. PERSIANN data are aggregated from 30 min rain estimations into hourly accumulated precipitations. In contrast, MODIS provides instantaneous observations twice a day. In this study, MODIS images within 20 min of PERSIANN estimations are mosaicked together into one raster image and then compared with corresponding PERSIANN data. Corresponding Stage IV data is presented for comparison of model performance. Figure 5.4 Panel (a) shows the Stage IV precipitation data (mm/h) on August 5th, 2007 (05:00 UTC). Panel (b) represents the corresponding PERSIANN data for the same time step (mm/h). By finding the ANN model's results on the corresponding MODIS images (two images for August 5th, 2007 (04:40 and 04:45 UTC)), the false alarms were identified. A false rain pixel is defined as a NR pixel in the ground-based observation data (Stage IV data) that contains precipitation from the satellite estimations (AghaKouchak and Mehran 2013). Figure 5.4c demonstrates the current algorithm's results in identifying false alarms on PERSIANN-derived precipitation. Grey pixels on the image show the location of correct rain detection from the satellite, and red and blue pixels are false rainy pixels from PERSIANN estimations. The blue color identifies the accuracy of the model in identifying NR pixels, while the red color demonstrates a false rain pixel that the model could not detect (here to define a false rain, the Stage IV data is considered the reference). Table 5.2 presents the number of rainy pixels in each dataset as well as number of FAR pixels detected. The algorithm was able to identify 155 false rain pixels (i.e. 62 % reduction in FAR for this event). Note that the region between the solid blue lines shows the MODIS coverage.

Figure 5.5 is another example of false rain detection for November 6th, 2007 (03:00 UTC). Panel (c) in the figure shows that the accuracy of the model is about 61 % in this event. PERSIANN estimation shows a large area of false rain on the southeast side of the event and the majority of FAR pixels could be removed using the current algorithm. Table 5.2 presents the number of rainy pixels in each dataset as well as number of FAR pixels detected.

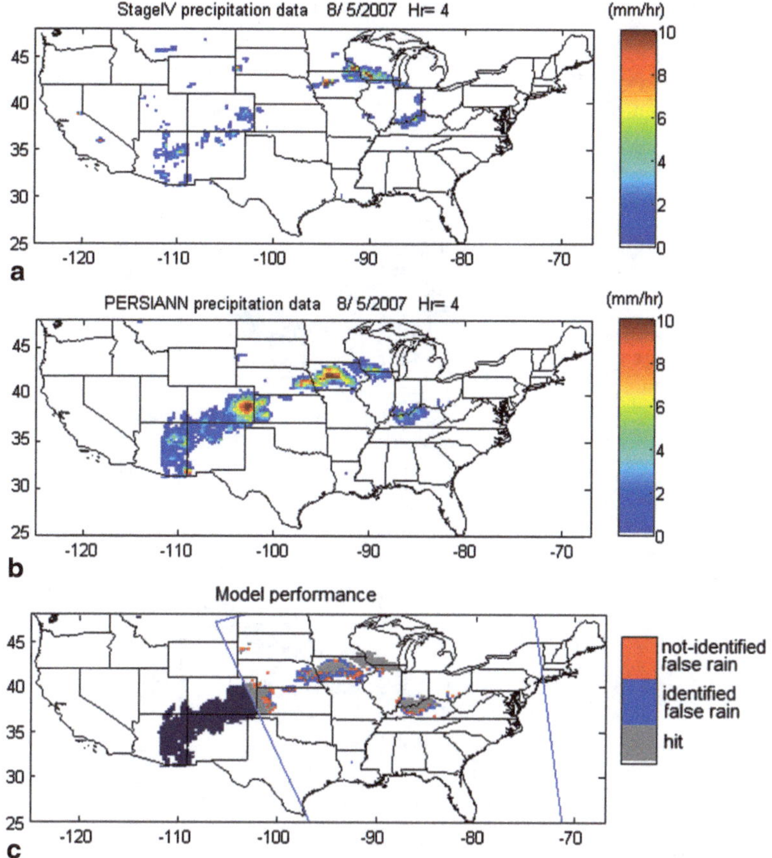

Fig. 5.4 Performance of the ANN model in identifying false rain locations on August 5th, 2007 (05:00 UTC). **a** Stage IV precipitation data (mm/h). **b** PERSIANN data (mm/h). **c** Model performance in FAR detection

5.7 Conclusion

Previous studies have highlighted the need to improve the quality of satellite precipitation data. High false alarm ratio is one of the problems that current satellite products are facing specially during cold seasons. In this study, the ability of a NR classification model using the CloudSat data as well as corresponding multispectral data from MODIS was investigated.

Table 5.2 Model performance presented in Figs. 5.4 and 5.5

	Summer	Winter
No. of precipitation pixels in the StageIV estimate	449	1151
No. of precipitation pixels in the PERSIANN estimate	562	717
No. of the false rain pixels corrected	155	300

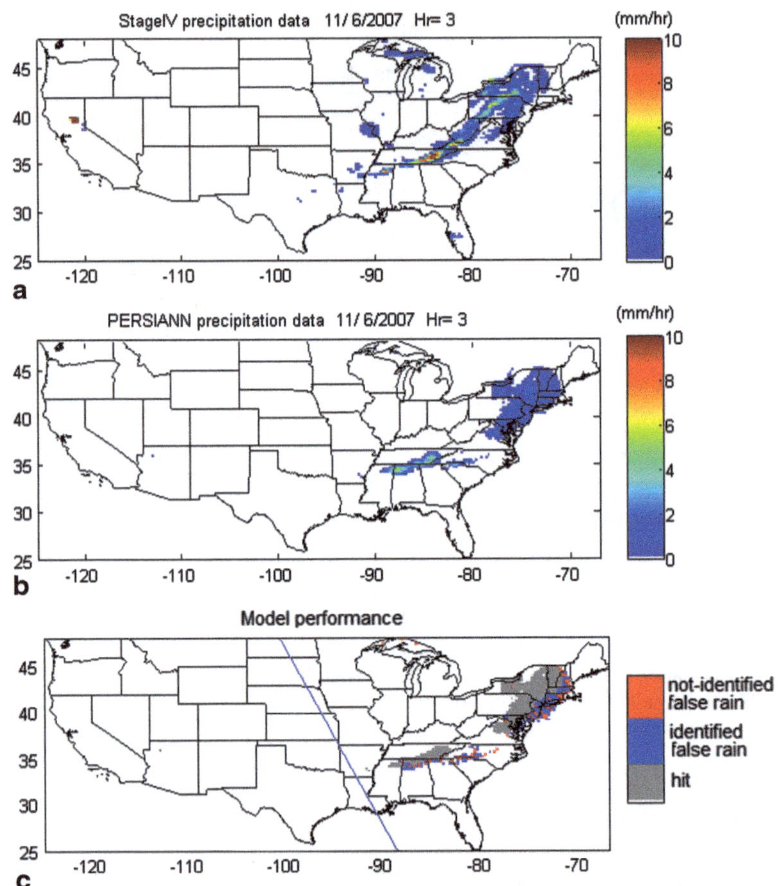

Fig. 5.5 Performance of the ANN model on November 6th, 2007 (03:00 UTC). **a** Stage IV precipitation data (mm/h). **b** PERSIANN data (mm/h). **c** Model performance in FAR detection

An artificial neural network model was developed to take advantage of accurate surface rain detections from the CloudSat satellite. The CPR radar data onboard CloudSat can detect the presence of hydrometeors near the surface. A separate training and validation dataset was considered to estimate the accuracy of the trained model. Model training was performed on CloudSat and MODIS data in the summer of 2008 and the winter of 2010. The summer and winter 2007 data sets were selected to assess the performance of the model. Model validation showed an accuracy of 93% for winter and 77% for summer in identifying false rain pixels. Different cloud classes were available from the CloudSat CLD-CLASS product, and the model performance was evaluated in presence of these different cloud classes. The model performance was the least accurate in case of deep convective and middle level (e.g. altostratus and altocumulus) cloud types.

By reducing false rain, the quality of satellite precipitation products for practical applications (e.g. flood forecasting) will significantly improve. In the future, it would be possible to include multi-spectral data from Advanced Baseline Imager (ABI) sensor aboard the future GOES-R satellite in order to overcome the limited retrievals of MODIS.

The proposed technique has the potential to be integrated into near real-time satellite precipitation data sets to reduce false alarms from the algorithms. Two case studies presented in the winter and summer 2007, using hourly PERSIANN data, showed a reduction of false rain in comparison with Stage IV radar data.

References

AghaKouchak, A., & Mehran, A. (2013). Extended Contingency Table: Performance Metrics for Satellite Observations and Climate Model Simulations, Water Resources Research, *49,* 7144–7149, doi:10.1002/wrcr.20498.

Haynes, J., L'Ecuyer, T., Vane, D., Stephens, G., & Reinke, D. (2011). Level 2-C Precipitation column algorithm product process description and interface control document. http://www.cloudsat.cira.colostate.edu/ICD/2C-PRECIP-COLUMN/2C-PRECIP-COLUMN_PDIC. Accessed 19 Oct 2014.

Sorooshian, S. (2011). Advancing the Remote Sensing of Precipitation, Bulletin of the American Meteorological Society—Nowcast, *92* (10), 1271–1272, doi:10.1175/BAMS-D-11-00116.1.

Chapter 6
Cloud Classification and its Application in Reducing False Rain

6.1 Introduction

Clouds are a key component in the weather and climate studies. However, their representations in climate models are associated with high uncertainty. For example, some studies show that compared to observations of real clouds, models significantly enhance solar radiation reflected by low clouds. This finding has major implications for the cloud-climate feedback problem in models (Stephens et al. 2008; Stephens 2010). A cloud classification scheme would be a valuable tool for illuminating the uncertainty of our models and algorithms and improving the accuracy of weather, climate, and precipitation studies. After classifying clouds into different classes, the precipitation estimation can be improved by integrating the classification scheme into the precipitation algorithm.

Different cloud classification techniques can be either statistically or physically based, using different cloud textural, spectral, and physical features obtained from satellite observations (Rossow and Schiffer 1999; Tian et al. 2000, Welch et al. 1992; Luo et al. 1995; Tovinkere et al. 1993; Bankert 1994; Wang and 2001; Bankert and Wade 2007). Physically based cloud type identification using weather satellites evolved during the 1980s and early 1990s mainly by using multi-spectral channel differences. The brightness temperature differences (BTD) between two or three channels were considered to identify a certain type of clouds. In addition to BTD, VIS and IR Channel combinations help to identify different cloud phases such as liquid, ice or mixed phase clouds (see Sect. 4.1 for details).

More sophisticated techniques include cloud microphysical and physical characteristics. For example, the International Satellite Cloud Climatology Project (ISCCP) (Rossow and Schiffer 1999) uses the information on cloud top pressure and cloud optical depth to classify clouds into seven groups: cumulus (Cu), stratocumulus (Sc), altocumulus (Ac), altostratus (As), nimbostratus (Ns), cirrus, cirrostratus or deep convective clouds.

In contrast, statistical classification methods can be a more effective means of including multichannel data and information to identify different cloud types under various surfaces and latitudes. Supervised and unsupervised statistical classification

N. Nasrollahi, *Improving Infrared-Based Precipitation Retrieval Algorithms Using Multi-Spectral Satellite Imagery,* Springer Theses, DOI 10.1007/978-3-319-12081-2_6

techniques such as the K-mean, Maximum Likelihood and Artificial Neural Network (ANN) have been used in multi-spectral image classification (Falcone and Azimi-Sadjadi 2005). In a supervised training of a model, a set of observations with their "true" cloud classifications is assigned and after a training period, this model predicts the cloud class for unknown cloud scenarios.

Hong et al. (2004) showed application of a feature-based cloud classification technique in satellite precipitation estimation. They classified clouds into a matrix of 20×20 based on cloud's coldness, texture and geometry. Then, they assigned a rain rate to each pixel based on the brightness temperature-rain rate relationship for each group. The results of their technique show promising results in incorporating cloud data into precipitation algorithms.

In this study a *cloud type* classification algorithm is developed to distinguish different clouds based on their multi-spectral features and CloudSat observations. After explaining the methodology and data, validation and application of the model is presented.

6.2 MODIS Cloud Mask

MODIS, a key instrument on NASA's EOS Terra and Aqua satellites, provides a cloud classification scheme (Cloud Mask; Platnick et al. 2003). The MODIS cloud classification takes advantage of three datasets including radiances in VIS, near infrared and IR images and BTD and texture (local standard deviation) of images. First, the mask identifies the likelihood of cloud cover for any given pixel by considering the reflectance in multi-spectral bands. The next step is identifying cloud top pressure by either a CO_2 slicing technique or emission from 11 μm channel. The third step is to determine the cloud's thermodynamic phase, and the last step retrieves optical thickness and particle size.

There are a total of 15 cloud classes in the MODIS cloud mask (presented in Table 6.1).

However, MODIS's physical cloud classification methods suffer from clouds' high variability and the dependence of cloud radiance on the emissivity of the surface over land. In addition, MODIS classification does not identify all cloud types, only cirrus and high clouds.

The cloud profiling radar onboard CloudSat can penetrate deeply into nearly all cases of non-precipitating clouds. Using a CloudSat profile radar cloud map, a cloud classification model can be trained and then used for better rainfall estimation.

6.3 Image Classification Using Self Organizing Maps

A nonlinear mapping Artificial Neural Network (ANN) system is developed to classify cloud images into seven cloud categories using CloudSat and MODIS data sets. The ANN architecture to be employed in this study is known as a Self-Organizing

Table 6.1 Initial classes from MODIS cloud mask

Class index	Content
1	Confident clear water
2	Confident clear coastal
3	Confident clear desert or semiarid ecosystems
4	Confident clear land
5	Confident clear snow or ice
6	Shadow of cloud or other clear
7	Other confident clear
8	Cirrus detected by solar bands
9	Cirrus detected by infrared bands
10	High clouds detected by CO_2 bands
11	High clouds detected by 6.7 mm band
12	High clouds detected by 1.38 mm band
13	High clouds detected by 3.7-and 12 mm bands
14	Other clouds or possible clouds
15	Others

Feature Map (SOFM) network (Kohonen 2006). The CloudSat radar images show distinguishable features of different cloud types. The classification layer categorizes MODIS images into a number of characteristic groups, each of which represents a specific cloud pattern in part of the input domain.

SOFM is an unsupervised classification technique that represents multidimensional input data in a lower dimensional space. In this study the input data are mapped into a 2D dimensional space. SOFMs are also considered a dimension reduction algorithm called vector quantization. Figure 6.1 shows the structure of a SOFM model. Inputs are fully connected to a two dimensional discrete map consisting of hexagonal nodes. Each vector of data from the input is placed onto one of the

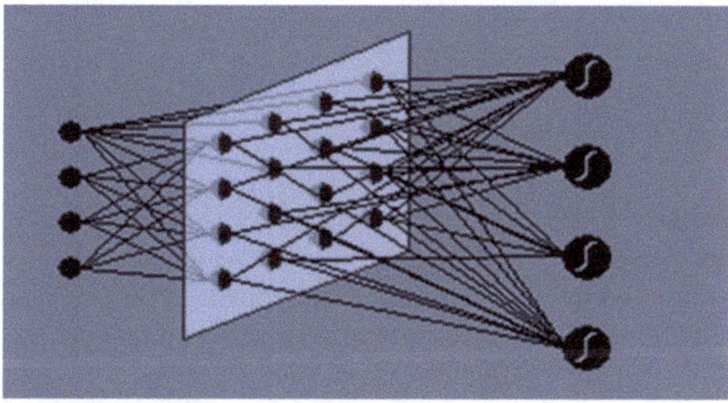

Fig. 6.1 Structure of a SOFM model

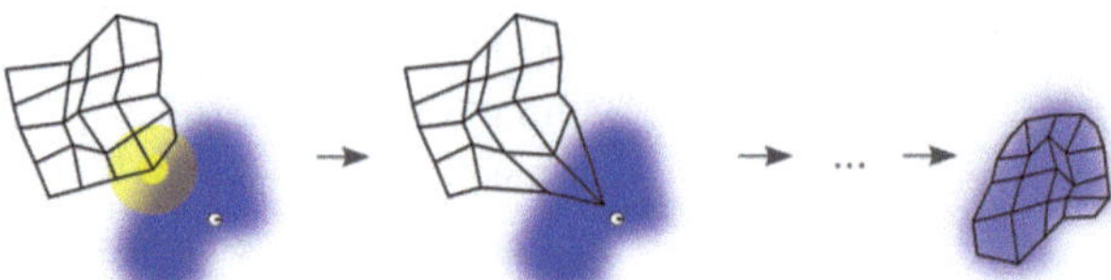

Fig. 6.2 A schematic diagram of the SOFM training (Source: http://en.wikipedia.org/wiki/Self-organizing_map)

grids of the map with the minimum distance between this vector and the map grid (closest weight vector).

Figure 6.2 schematically shows how the model structure is trained. The blue area shows the input space (training dataset). The SOFM nodes (black grids) are randomly distributed in the space. The white dot shows the current training vector from the training dataset. The nodes closest to the input vector (highlighted in yellow) will be moved to have a minimum distance to the training vector. After introducing all the vectors in the training dataset to the model, the final map will be a representative of input data distribution. During the training, the distance between the input vector (x_i) and the node centers will be calculated (Equation 6.1)

$$d_j = \left[\sum_{i=1}^{n_1} (x_i - \omega_{ij})^2 \right]^{\frac{1}{2}}; \qquad where \quad j = 1, \ldots, n \tag{6.1}$$

the best-matching SOFM cluster c (winning node) is the one corresponding with the minimum distance (d_c) between the input feature vector and the SOFM connection weights ω_{ij} (Equation 6.2). ω_{ij} is the weight matrix that represents the center of clusters.

$$\mathrm{d_c} = \min(\mathrm{d_j}); \qquad \text{where} \quad \mathrm{j} = 1, \ldots, \mathrm{n} \tag{6.2}$$

A SOFM with 15×15 hexagonal nodes are applied in this study.

6.4 Data Pre-processing

Data Normalization Brightness temperature values and reflectance data have different units and ranges of variability. By normalizing there data, we reduce the effects of data range variations. Scaling all the values so that they fall in the range of [0—1] will improve the model's training accuracy and reduce the training time.

Visible Data Correction Visible data should be normalized based on the sun's direction and the time of day. Behrangi (2009) showed the effectiveness of normalizing visible data by the sun zenith angle (SZA). He concluded that multiplying by $\cos(SZA)^{-1}$ resulted in a larger portion of visible data that can be used in the analyses. This study uses the same approach to normalize the visible channels. Only pixels with SZA< 60° are considered to minimize the uncertainty associated with a large SZA.

$$Ref\_norm = Ref * \cos(SZA)^{-1} \tag{6.3}$$

Uniform Distribution of Data One of the other methods to increase the accuracy of model's training is to use a uniform distribution of data. Different cloud types have different distributions and that affects the results of the model outcome. The higher the number of samples in the training dataset, the more likely it is that model will be tuned toward a given specific cloud type (Hsu et al. 2002).

6.5 Training and Validation Datasets, Summer Season

The training does not include clear sky conditions because they happen more often than cloudy scenes and influence the ANN model weights (Capacci and Conway 2005). To filter out clear sky pixels, a cloud mask is applied first. To identify cloudy scenes, pixels with the highest probability of cloud presence from the MODIS cloud mask dataset are considered as cloudy pixels.

Figure 6.3 represents the variability in the input data. The red line in the middle of the box represents the median and two boundaries of the box are the twenty-fifth and seventy-fifth percentile of the data. The outliers are plotted separately by a red cross. By definition, the outlier is a value that is more than 1.5 times the interquartile range (length of the box) away from the top or bottom of the box. The outlier data plotted in the figure are included in the training of the model since they are representative of upper and lower tails of distribution corresponding to extreme observation data. There is a large difference between the range of values in water vapor and IR channels compared with visible data as shown in the figure. Normalizing the values of the IR and VIS channels, confining them to a range of 0—1 removes the effects of different units in the dataset. Visible data are also normalized to account for the effects of sun zenith angel using Equation 6.3.

Figures 6.4 and 6.5 present the distribution of data used for training and validation for summer 2008 and 2007, respectively. Summer 2008 training data consist of more than 121,000 cloudy pixels and summer 2007 covers 70,000 cloudy samples. Cloud classes of cumulus and nimbostratus in summer datasets have very limited samples (occurrence) and were removed from the analysis.

The figures also show that high clouds are the dominant type of clouds during summertime. Alto-stratus and altocumulus clouds have almost the same distribution. Cloud type distribution for 2007 and 2008 are quite similar. Note that distribution of different cloud types is not uniform. High clouds happen more often than any other cloud type. Using the original data distribution in the training algorithm will tune model parameters toward high clouds (dominant cloud type). As explained in Sect. 6.4, a uniform distribution of data is introduced to the model in the training phase.

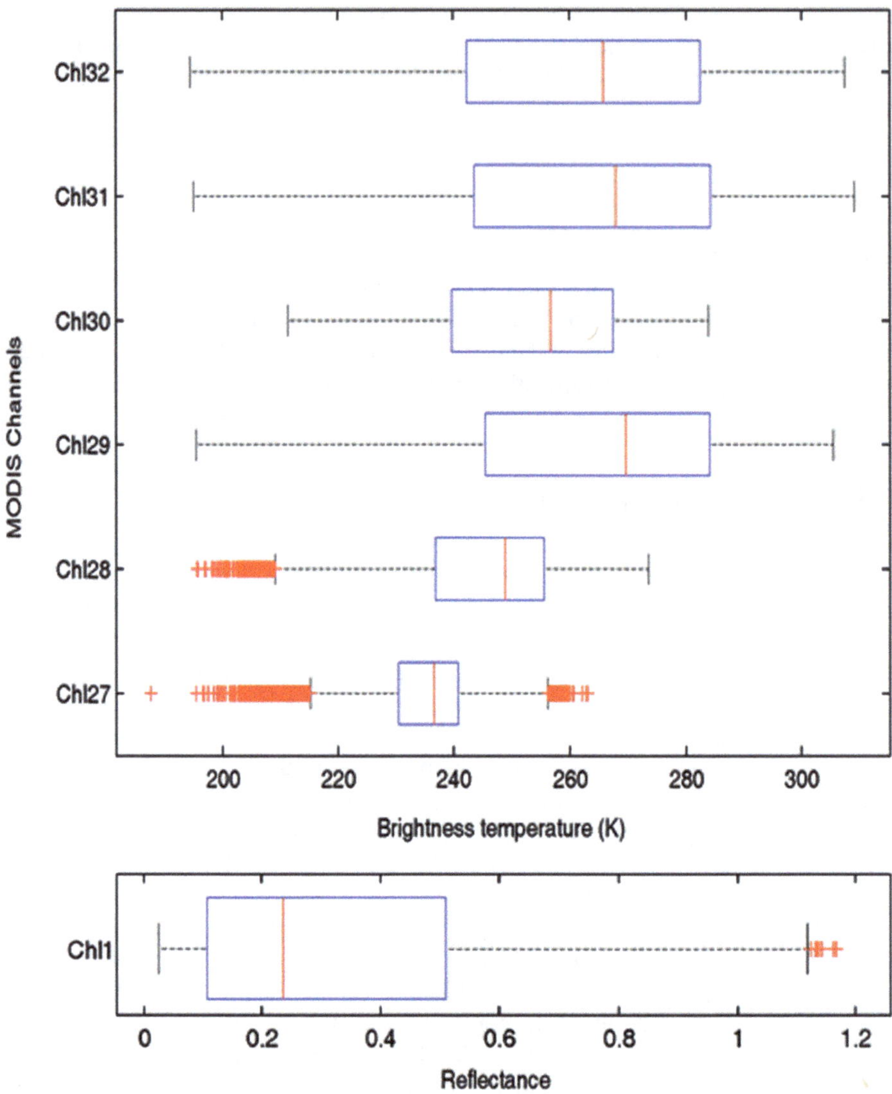

Fig. 6.3 A box-plot of 6 MODIS channel input data, summer 2008

6.6 Training and Validation Datasets, the Winter Season

Winter 2010 and 2007 data, each with more than 50,000 cloudy pixels, are considered in the training and validation process. Figures 6.6 and 6.7 show the histogram of 2010 and 2007 datasets, respectively. Comparing figures, large differences can be seen in cloud type distributions. Nimbostratus, a thick cloud that causes precipitation with prolonged rain events, was the dominant cloud type in winter 2010. On the other hand, winter 2007 was relatively drier and had mainly inhomogeneous

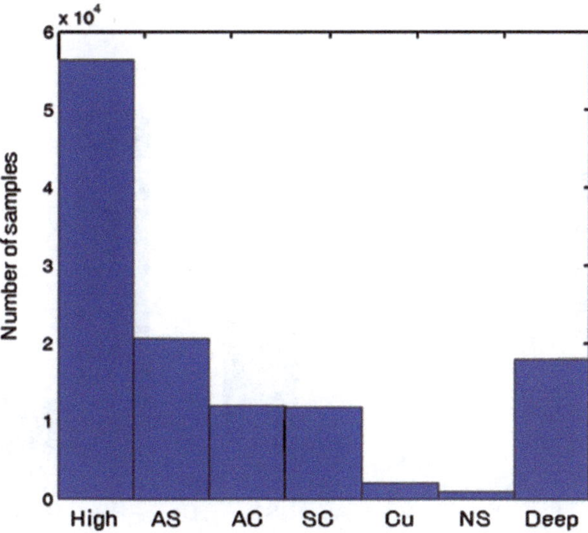

Fig. 6.4 Distribution of different cloud types in the input data, summer 2008

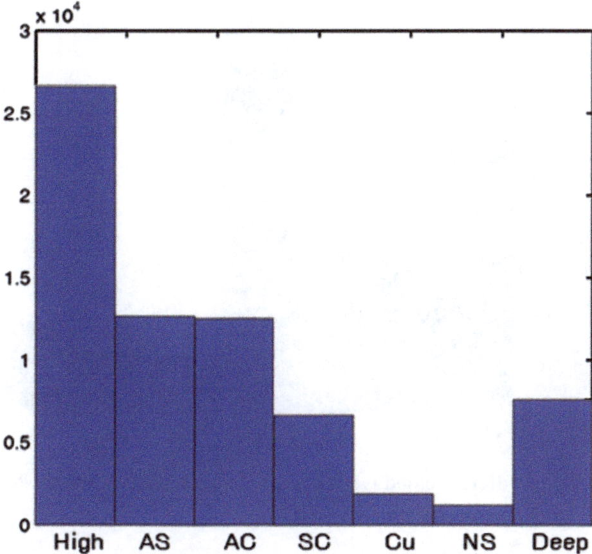

Fig. 6.5 Distribution of different cloud types in the validation dataset, summer 2007

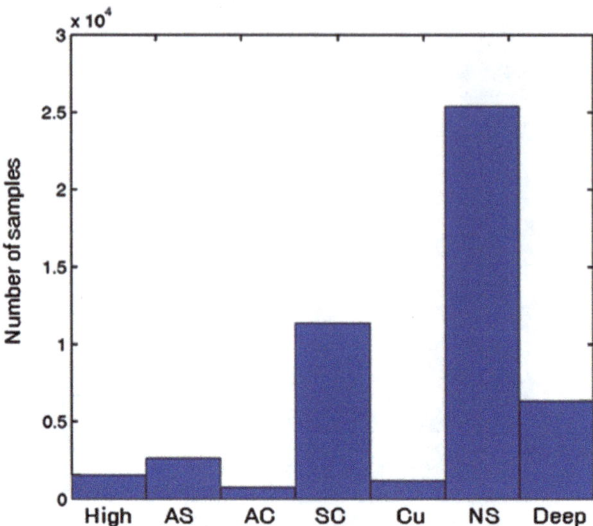

Fig. 6.6 Distribution of different cloud types in the training dataset, winter 2010

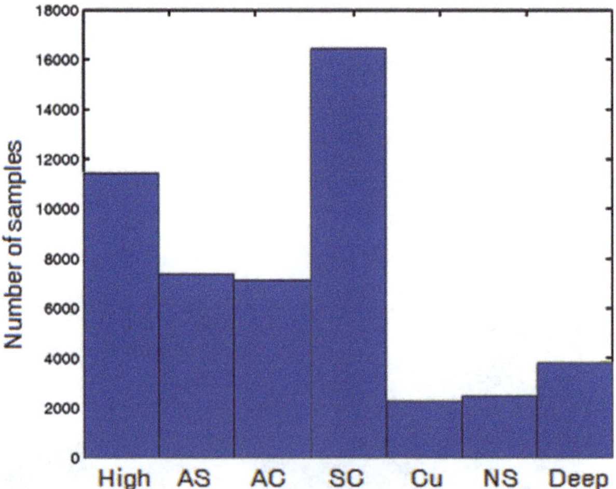

Fig. 6.7 Distribution of different cloud types in the validation dataset, winter 2007

shallow stratocumulus clouds. Many pixels were associated with no-rain clouds, such as high and altostratus clouds.

Because of large differences in the distribution of data in 2010 and 2007, and also to increase the sample size, both years' data are combined and a subset with uniform distribution is considered for training.

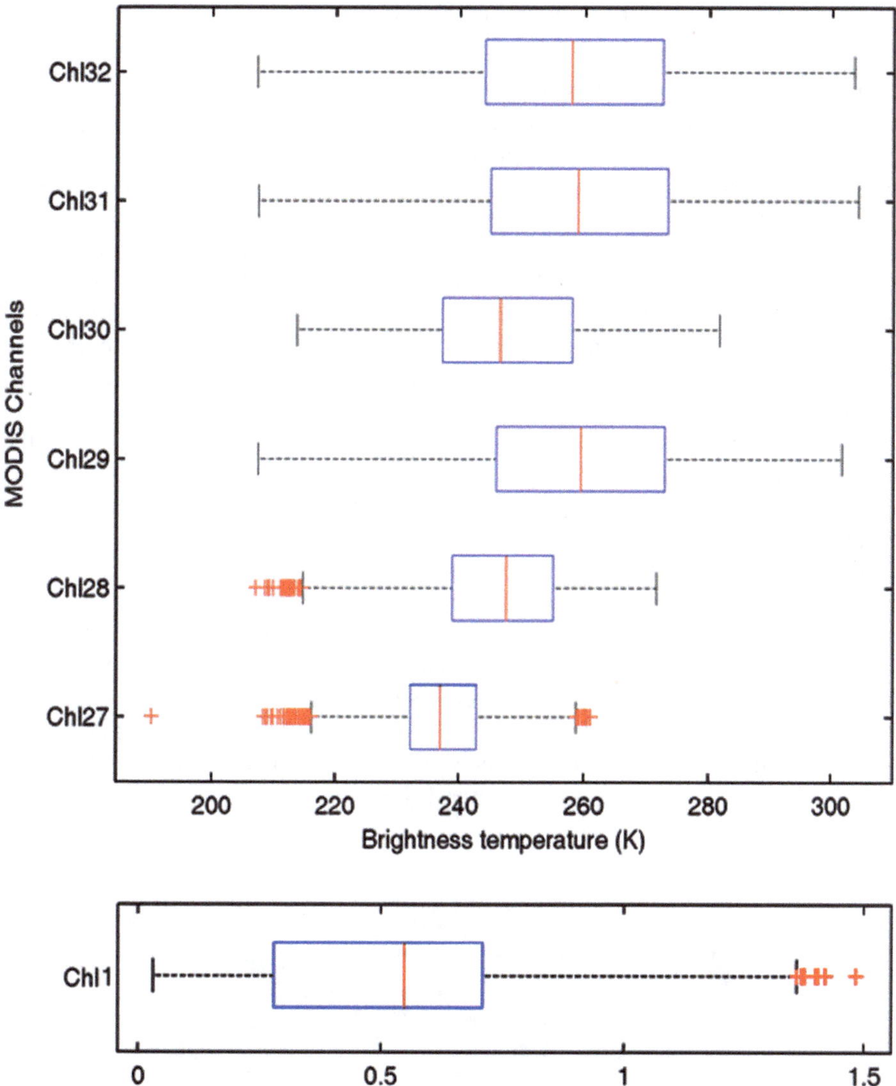

Fig. 6.8 A box-plot of 7 MODIS channel input data, winter 2010

Figure 6.8 shows the distribution of multi-spectral data for 7 MODIS spectral channels that are used in the training phase. As shown in the figure, channels 29, 31 and 32 have a similar range of data, and channel 27 (6.535–6.895 μm), a water vapor channel, has a slightly different range of brightness temperature values compared to other channels.

6.7 SOFM Model for Summer Season

Using a uniform distribution of input data, the SOFM model is trained to classify clouds into different groups. To select a uniform distribution, the cloud type with minimum occurrence was selected and the same number of pixels was randomly picked from other cloud types in the training dataset. In the summer dataset, deep clouds occur less often and the same number of cloudy pixels was chosen from other cloud groups. After about 5000 iterations, the data samples are distributed onto a 15×15 map. Figure 6.9 shows the sample distribution on a 2-D map. The figure also shows the structure of the SOFM nodes. Each node has a hexagonal shape connected to six neighboring nodes. The number in the middle of each node shows the number of samples located on each node. For example, a node labeled 178 represents 178 cloudy pixels from the training dataset. Pixels are arranged in correspondence to other pixels with similar properties. There are two areas on the map with larger distributions of samples, one on the right and the second one on the left. The size of the hexagons on the map is related to the number of samples on that particular node.

Figure 6.10 shows the weight (cluster center) of each input feature on the SOFM map. The normalized values (ranging from 0 to 1) are shown with corresponding colors changing from blue to red. Blue represents smaller values (colder brightness temperature or lower reflectance on VIS data) and red represents warmer pixels or pixels with higher reflectance. Comparing subplots of Fig. 6.10 shows the clusters located on the lower right corner of the map correspond to colder pixels with lower brightness temperature (higher elevated clouds) and pixels on the bottom and left corners have higher reflectance values (most likely thicker clouds). Comparing

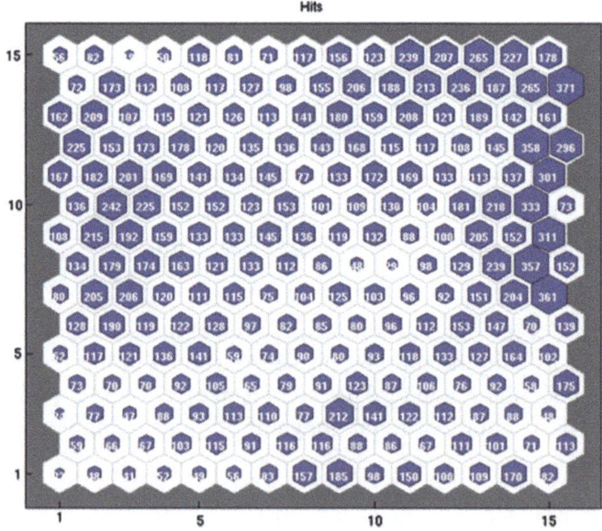

Fig. 6.9 Distribution of different samples on a SOFM map

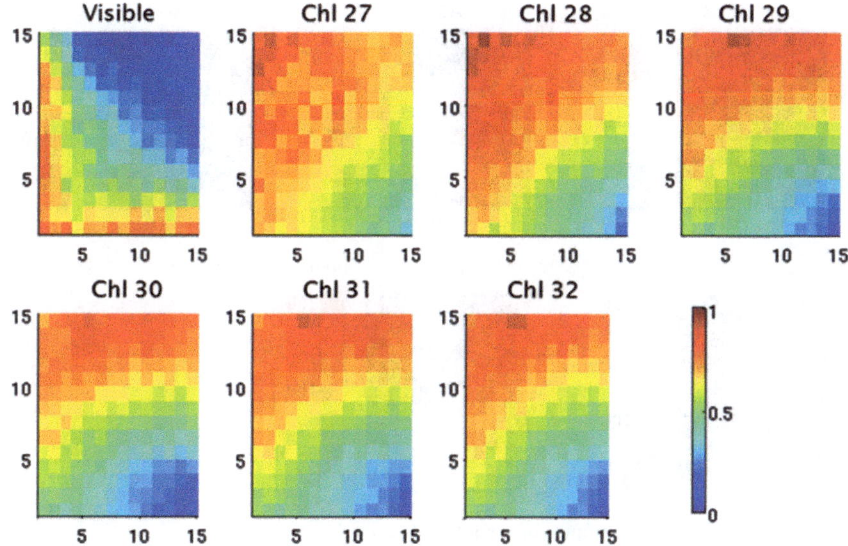

Fig. 6.10 Weight of different input features on the SOFM map

Figs. 6.9 and 6.10 reveals that most of the samples are located on the clusters with low reflectance and medium temperature on IR (clusters on the right hand side) or warmer tops with higher reflectance (left hand side).

After training the algorithm, we can check the distribution of cloud pixels on the SOFM map. Fig. 6.11 shows the number of cloudy pixels on each cluster. Comparing Figs. 6.11 and 6.10 shows some general characteristics of clouds. For example, deep convective clouds are positioned on the lower part of the map. This area corresponds to low brightness temperature and high reflectance in the visible channel. Deep convective clouds have the coldest top temperatures. As explained earlier, the total number of samples selected from each cloud group in the training dataset is the same.

Figure 6.12 shows the probability of each cloud type on the SOFM clusters. Equation 6.4 explains how to calculate the probability of each cloud type on each node.

$$P(x_{ij}) = \frac{S_{ij}}{\sum_{k=1}^{N} S_{kj}} \tag{6.4}$$

Where, $P(x_{ij})$ is the probability of cloud type i on cluster j.

S_{ij} is the number of cloud samples of type i on cluster j.i changes from 1 to 7 (7 cloud types) and j changes from 1 to 225 (total number of clusters on a 15×15 map). S_{kj} represents the total number of cloud samples on node j.

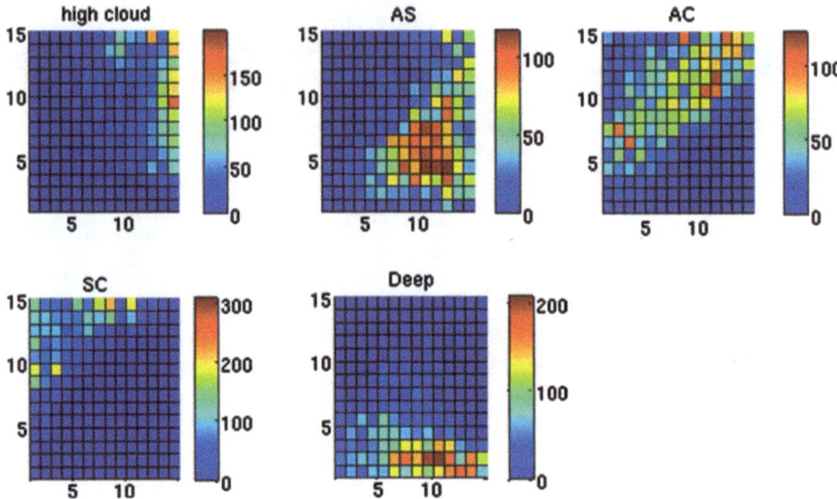

Fig. 6.11 Number of cloud samples on each SOFM cluster

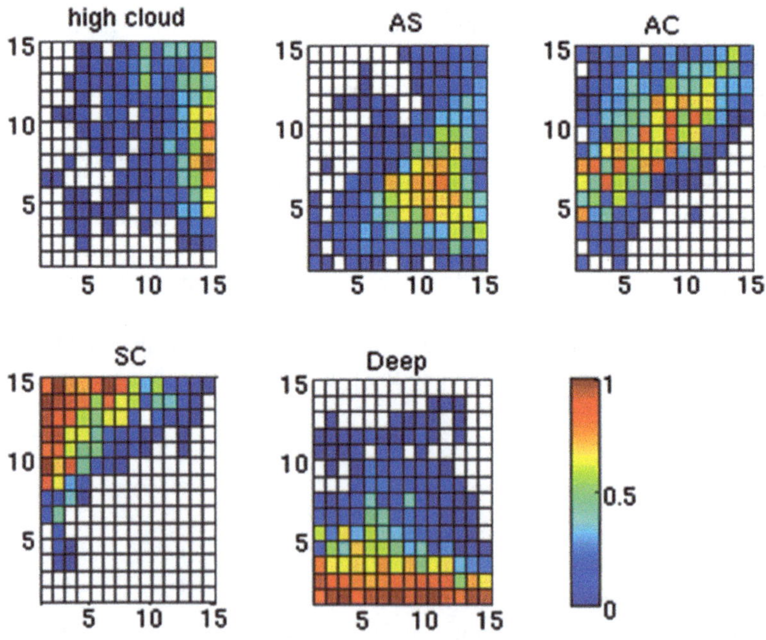

Fig. 6.12 Probability of various cloud types on SOFM clusters

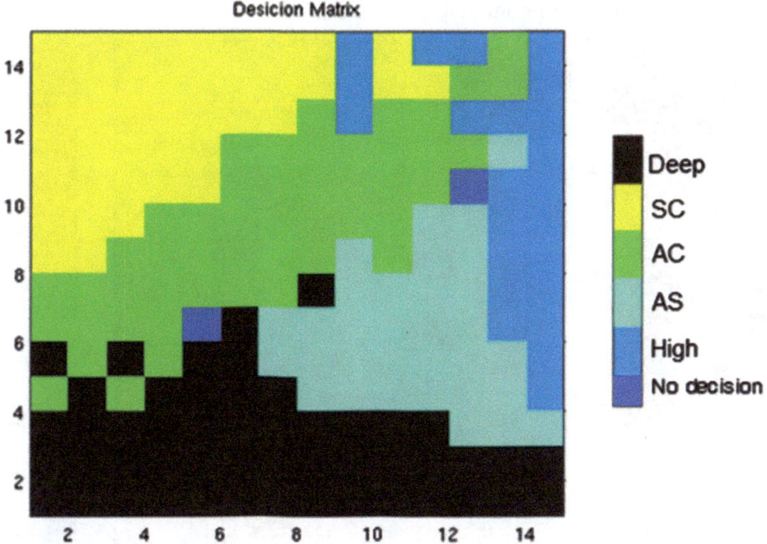

Fig. 6.13 Decision matrix, summer season

After finding the probability of each class on every cluster, the dominant cloud class on each cluster can be determined. The dominant cloud type is the cloud type with the highest probability. After finding the most probable cloud type at each cluster, the decision matrix can be generated. If there is no cloud sample located on any particular cluster, there is no decision on that cluster. Figure 6.13 represents the decision matrix for the summer season. There are two no-decision clusters in the summer season decision map.

The confidence of each decision cluster can also be determined. The confidence is defined as the probability of the dominant cloud type on each cluster. The higher the probability, the better the confidence level of the classification.

Figure 6.14 shows the confidence level of the classification map. Comparing Figs. 6.14 and 6.13 shows higher confidence in occurrence of high cloud and stratocumulus. The decision confidence on the upper right corner clusters is low. The upper right corner corresponds to low reflectance on the visible channel and warm tops (mainly middle level clouds).

6.8 Validation of Cloud Classification Model, Summertime

To demonstrate the accuracy of the classification model, the model is calibrated against the summer 2007 dataset. Knowing the original cloud types from CloudSat, the 2007 pixels were classified using the created model and then compared with the

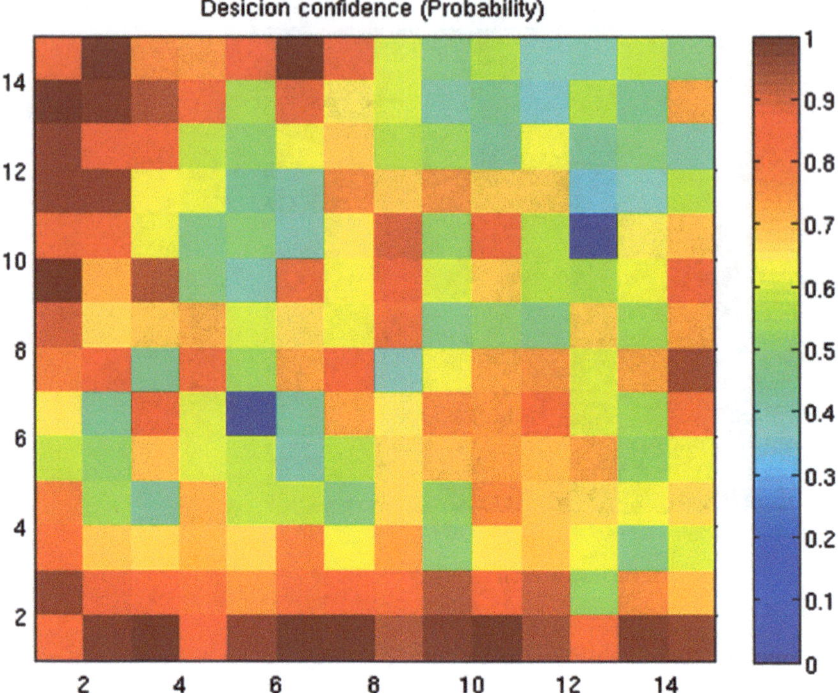

Fig. 6.14 Decision confidence of summer classification

CloudSat cloud types. The results are presented in Fig. 6.15. High clouds and deep convective clouds have the highest accuracy and the classification accuracy is lowest in the case of middle level clouds. As discussed earlier the miss-classification between high cirrus and deep convective is one of the shortcomings of IR-based algorithms that leads to false rain detection. Integrating the current cloud classification model into the precipitation algorithm, reduces the false rain in presence of high clouds.

6.9 SOFM Model for the Winter Season

A uniform distribution of data drawn from winter 2010 and 2007 data is considered to train the SOFM model. After 5000 iterations, the distribution of samples on the SOFM map is presented in Fig. 6.16.

Figure 6.17 shows the weight (cluster center) of each input feature on the SOFM map in winter classification data. The X and Y axes are the 15 × 15 cluster map and the normalized brightness temperature or visible reflectance data (ranging from zero to one) are shown on the map. The left hand side and bottom clusters, correspond to high visibility and the upper right corner corresponds to pixels with lower

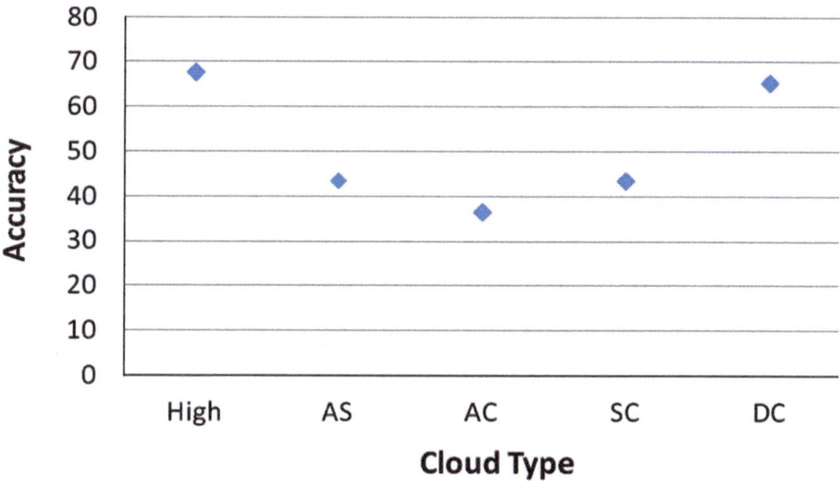

Fig. 6.15 The accuracy of summertime cloud classification

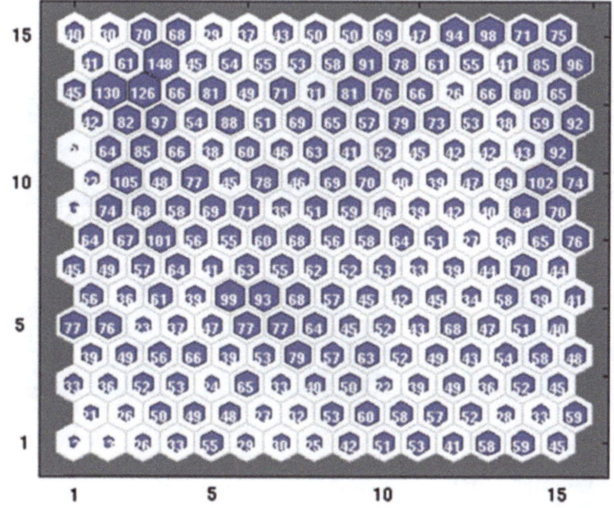

Fig. 6.16 Distribution of training samples on the SOFM map

visibility. Higher brightness temperature samples are placed on the top part of the SOFM map (shown in red on Fig. 6.17).

Figure 6.18 shows the distribution of cloudy samples from each cloud type on the 2D map. Because we have used a uniform distribution of clouds in the training (Hsu et al. 2002), one can see that there are equal numbers of samples from each cloud type distributed on the SOFM clusters.

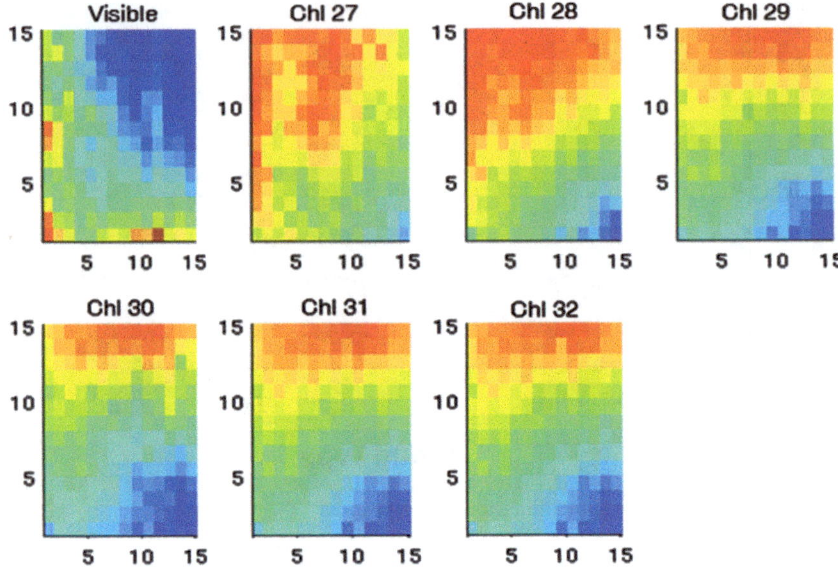

Fig. 6.17 Weight of different input features on the SOFM map

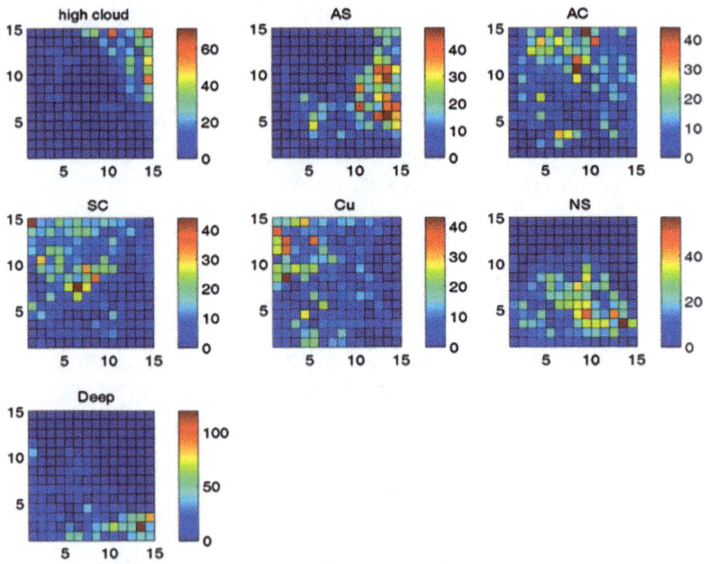

Fig. 6.18 Number of cloud samples on each SOFM cluster

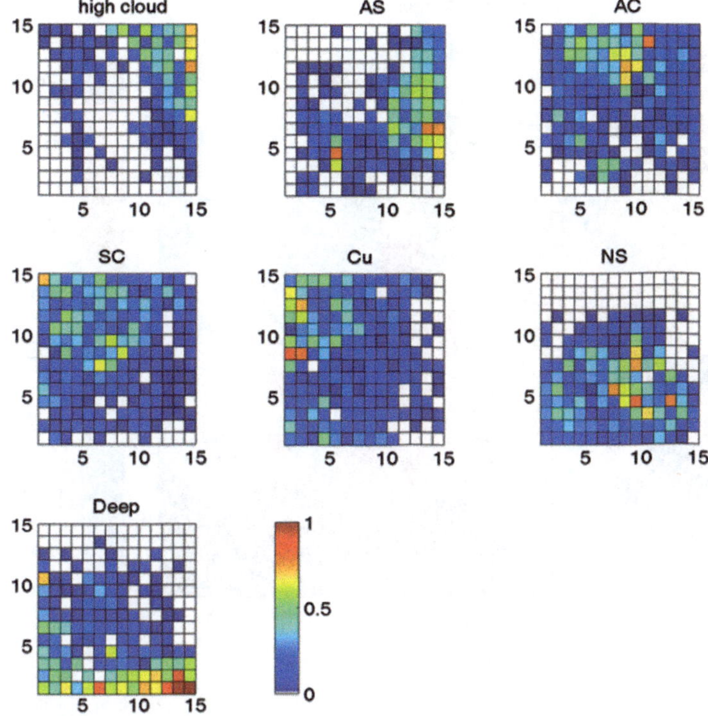

Fig. 6.19 Probability of various cloud types on SOFM clusters

Deep clouds, such as deep convective clouds, appear on the lower part of the map that corresponds to low brightness temperature and high visible reflectance. In general, clouds with high albedo have a large optical depth and are thicker. On the other hand, high clouds correspond to low reflectance due to their shallow depth. Middle level clouds fall in the middle of the 2D map. NS clouds are another distinct type of cloud that are located near deep convective clouds on the SOFM map but appear lower in the atmosphere (warmer in IR channels) and have lower albedo.

Equation 6.4 is used to find the probability of each cloud type on each cluster. Fig. 6.19 shows the probability of each cloud type on every cluster. The figure shows low probability of cloud types and spread distribution of samples.

The decision matrix depicts the dominant cloud type on each cluster. After running the model three times, the decision matrix is created based on the results and is presented in Fig. 6.20.

Figure 6.21 shows the decision confidence for the winter season classification. The low confidence values represent poor performance of the model in the winter season. In general, satellite observations have better results in summer compared to the winter due to dominance of convective storms. (see also, Mehran and AghaKouchak 2014; Sorooshian et al. 2011; AghaKouchak and Mehran 2013)

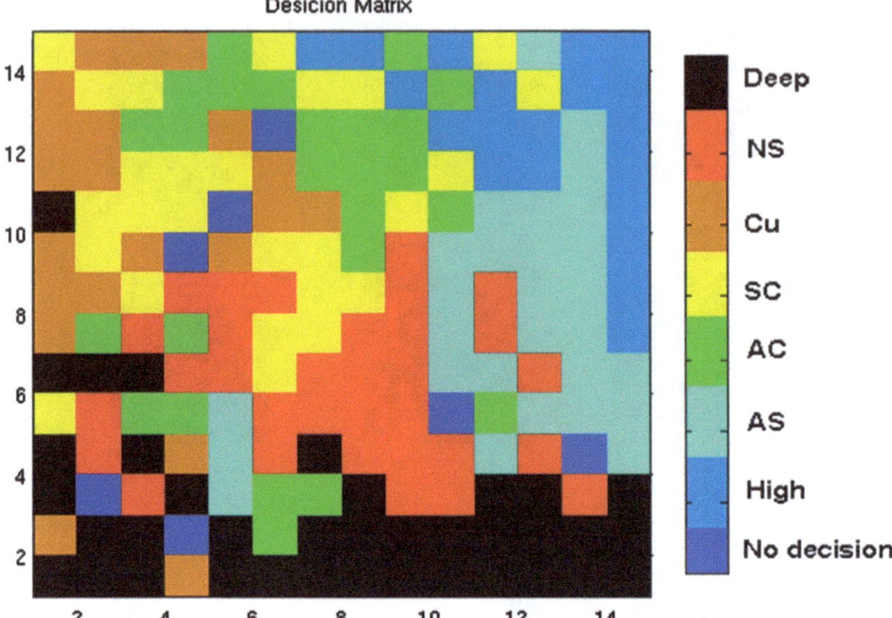

Fig. 6.20 Decision matrix, the winter season

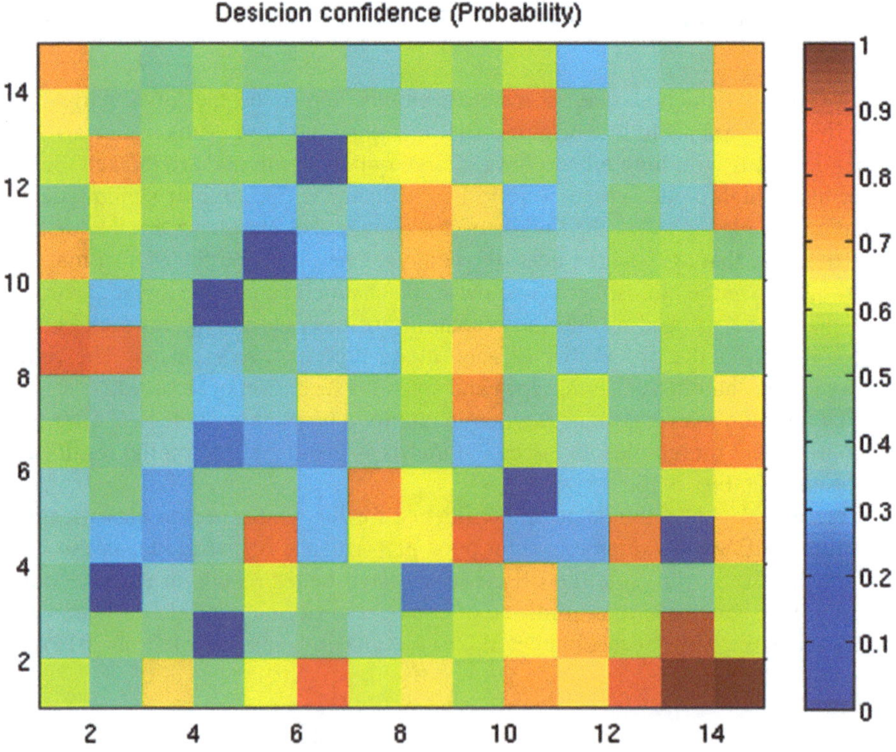

Fig. 6.21 Decision confidence of winter classification

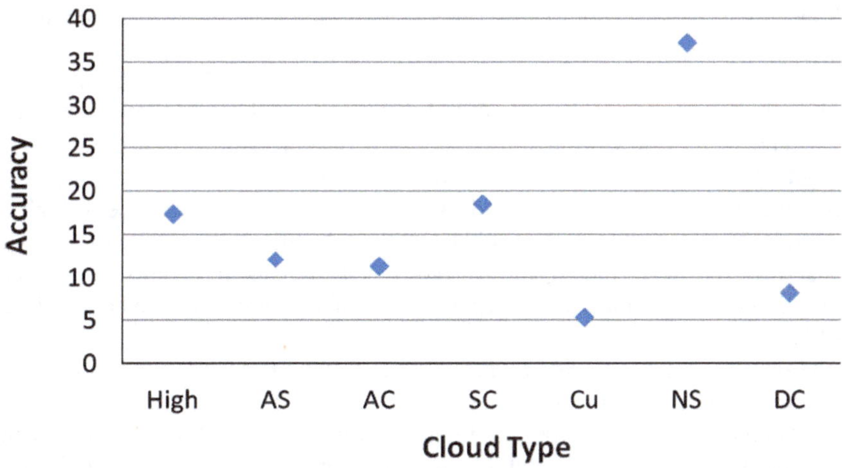

Fig. 6.22 The accuracy of winter season cloud classification

6.10 Validation of Cloud Classification Model, the Winter Season

Validation of the winter season dataset is performed on a subset of the 2007 and 2010 dataset that was not used in the model training. The validation results show low accuracy in the classification of cloud groups. Nimbostratus clouds have the highest accuracy that is 37.2 % (Fig. 6.22).

6.11 Conclusion

The poor result of cloud classification in the winter season is not surprising because satellite observations tend to have weaker observation accuracies in winter. Ackerman et al. (2009) confirm discrepancies among different PMW sensors in detecting wintertime high latitude cloud properties because of similar spectral radiances between clouds and the background area. The same study highlights differences between MODIS sensors onboard Terra and Aqua due to their respective instrument performances.

The shortcomings of wintertime satellite observations are not unique to cloud detection. Aghakouchak et al. (2012) discusses high systematic errors in wintertime satellite observations in detecting precipitation. Consistently missed precipitation in satellite products was also the dominant cause of error in winter observations as reported by Tian et al. (2009). McCollum et al. (2002) studied PMW-based satellite rainfall data over the United States and showed underestimations in wintertime. In addition, Rozumalski (2000) showed better IR-based satellite estimations in summertime compared to winter season.

The main reason in better summertime estimation is the dominance of convective clouds. Because of their high altitude and ice particles, convective clouds are easier to be detected in satellite rain estimations. The probability of detection of rain is very high in summer estimations as shown in Chap. 2.

Summertime distribution of clouds also shows significant presence of high cirrus clouds. The base of high clouds is located in elevations higher than 7 km. They consist of ice particles and appear very cold in IR images. In contrast to deep convective clouds, high clouds are thin. Multi-spectral data are shown to be effective in identifying cirrus clouds.

The distribution of clouds in wintertime is different. A large portion of cloudy pixels are associated with nimbostratus (winter 2010) and stratocumulus (winter 2007). Both of these cloud types are low to mid-level clouds. Ns clouds are located at elevations lower than 4 km and Sc clouds appear lower than 2 km altitude. Their low altitude makes them appear warm in the IR brightness temperature data and challenging to be identified. Ns clouds sometime look similar to other middle level clouds such as stratus and stratocumulus or even altostratus clouds. The difference between Ns and the three mentioned clouds is that Ns clouds produce precipitation.

In addition, as discussed in Sect. 3.2, wintertime detection of middle-level clouds in cold winters in high and mid-latitudes is challenging. The snow-covered cold surface might not be distinguishable from low and middle-level clouds. It is also difficult to identify low or middle-level clouds in high mountainous regions since the surface elevation may be at the same elevation as clouds.

These results show that with the current sensor capabilities it is difficult to identify middle level clouds and there is a need for radar data. Radar can penetrate through the clouds to see the vertical structure of clouds. Having a three dimensional observation, one can provide a cloud classification algorithm with the ability to identify middle and low level clouds in addition to high level clouds.

References

Ackerman, S., Frey, R., & Holz, R. (2009). The challenge of passive remote sensing of clouds from satellites over polar regions in winter. Fifth annual symposium on future operational environmental satellite systems- NPOESS and GOES-R. https://ams.confex.com/ams/89annual/techprogram/paper_150182.htm

AghaKouchak, A., Mehran, A., Norouzi, H., & Behrangi, A. (2012). Systematic and random error components in satellite precipitation data sets. *Geophysical Research Letters, 39,* L09406.

AghaKouchak, A., & Mehran, A. (2013). Extended Contingency Table: Performance Metrics for Satellite Observations and Climate Model Simulations, Water Resources Research, *49,* 7144–7149, doi:10.1002/wrcr.20498.

Bankert, R. L. (1994). Cloud classification of AVHRR imagery in maritime regions using a probabilistic neural network. *Journal of Applied Meteorology, 33,* 909–918.

Bankert, R. L., & Wade, R. H. (2007). Optimization of an instance-based GOES cloud classification algorithm. *Journal of Applied Meteorology and Climatology*, *46,* 36–49.

Behrangi, A., Hsu, K., Imam, B., Sorooshian, S., & Kuligowski, R. (2009). Evaluating the utility of multi-spectral information in delineating the areal extent of precipitation. *Journal of Hydrometeorology, 10*(3), 684–700.

Capacci, D., & Conway, B. (2005). Delineation of precipitation areas from MODIS visible and infrared imagery with artificial neural networks. *Meteorological Applications, 12,* 291–305.

Falcone, A. K., & Azimi-Sadjadi, M. R. (2005). Multi-satellite cloud product generation over land and ocean using canonical coordinate features. OCEANS, 2005. *Proceedings of MTS/IEEE OCEANS* (Vol. 1. pp. 638–644). Piscataway: IEEE.

Hong, K. L., Hsu, S. S., & Gao, X. (2004). Precipitation Estimation from Remotely Sensed Imagery Using an Artificial Neural Network Cloud Classification System.J. Appl. Meteorol, *43,* 1834–1852.

Hsu, K., Gupta, H., Gao, X., Sorooshian, S., & Imam, B. (2002). Self-organizing linear output map (SOLO): An artificial neural network suitable for hydrologic modeling and analysis. *Water Resources Research, 38*(12), 1–17.

Kohonen, T. (2006). Self-organizing neural projections. Neural Networks, *19*(6), 723–733.

Luo, G., Davis, P. A., Stowe, L. L., & McClain, E. P. (1995). A pixel-scale algorithm of cloud type, layer, and amount for AVHRR data. Part I: Nighttime. *Journal of Atmospheric and Oceanic Technology, 12,* 1013–1037.

McCollum, J. R., Krajewski, W. F., Ferraro, R. R. & Ba, M. B. (2002). Evaluation of biases of satellite rainfall estimation algorithms over the continental United States. *Journal of Applied Meteorology, 41,* 1065–1080.

Mehran, A., AghaKouchak, A. (2014). Capabilities of Satellite Precipitation datasets to Estimate Heavy Precipitation Rates at Different Temporal Accumulations, Hydrological Processes, *28,* 2262–2270, doi:10.1002/hyp.9779.

Platnick, S. A., King, M. D., Ackerman, S. A., Menzel, W. P., Baum, B. A., Riédi, J. C., & Frey, R. A. (2003). The MODIS cloud products: Algorithms and examples from Terra. *IEEE Transactions on Geoscience and Remote Sensing, 41,* 459–473. doi: 10.1109/TGRS.2002.808301.

Rossow, W. B., & Schiffer, R. A. (1999). Advances in understanding clouds from ISCCP. *Bulletin of the American Meteorological Society, 80,* 2261–2286.

Rozumalski, R. A. (2000). A quantitative assessment of the NESDIS Auto-Estimator. *Weather and Forecasting, 15,* 397–415.

Sorooshian, S., AghaKouchak, A., Arkin, P., Eylander, J., Foufoula-Georgiou, E., Harmon, R., Hendrickx, J., Imam, B., Kuligowski, R., Skahill, B., Skofronick-Jackson, G. (2011). Advanced Concepts on Remote Sensing of Precipitation at Multiple Scales, Bulletin of the American Meteorological Society, *92* (10), 1353–1357, doi: 10.1175/2011BAMS3158.1.

Stephens, G. (2010). Is there a missing low cloud feedback in current climate models? *GEWEX News, 20*(1), 5–7.

Stephens, G. L., et al. (2008). CloudSat mission: Performance and early science after the first year of operation. *Journal of Geophysical Research, 113,* D00A18. doi:10.1029/2008jd009982.

Tian, B., Azimi-Sadjadi, M. R., Vonder Haar, T. H., & Reinke, D. (2000). Temporal updating scheme for probabilistic neural network with application to satellite cloud classification. *IEEE Transactions on Neural Networks, 11,* 903–920.

Tovinkere, V. R., Penaloza, M., Logar, A., Lee, J., Weger, R. C., Berendes, T. A., & Welch, R. M. (1993). An intercomparison of artificial intelligence approaches for polar scene identification. *Journal of Geophysical Research, 98,* 5001–5016.

Wang, Z., & Sassen, K. (2001). Cloud type and macrophysical property retrieval using multiple remote sensors. *Journal of Applied Meteorology, 40,* 1665–1682.

Welch, R. M., Sengupta, S. K., Goroch, A. K., Rabindra, P., Rangaraj, N., & Navar, M. S. (1992). Polar cloud and surface classification using AVHRR imagery: An intercomparison of methods. *Journal of Applied Meteorology, 31,* 405–420.

Chapter 7
Summary and Conclusions

False alarm is one of the shortcomings of satellite precipitation estimates that needs to be improved. Many studies have quantified the FAR, bias and errors of satellite precipitation estimates. However, reducing the FAR is an essential step in improving the quality of satellite data. In this research, three techniques are proposed to reduce the FAR by integrating information from multi-spectral satellite imagery as well as satellite radar observations. MODIS, a multi-spectral satellite sensor, observes the atmosphere in 36 spectral channels, providing a special source of information for cloud observation. On the other hand, CloudSat has two products, Cloud Type and Precipitation Occurrence, that can add a new dimension to the IR-based precipitation algorithms. In the first approach, the cloud type classification dataset from CloudSat was used as a reference to find the non-precipitating cloud types. One of the reasons for FAR in satellite precipitation data is the presence of high non-precipitating clouds such as cirrus or cirrus anvil. Generally, the areal coverage of satellite precipitation estimation is larger than that of ground observation, primarily due to presence of cirrus anvil. Finding the pixels with anvil coverage, one can eliminate the false rain estimations from the satellite product. A trained neural network model using six MODIS water vapor, window and infrared channels (6.75, 7.325, 8.55, 9.7, 11.03, 12.02 μm wavelength) as the input and CloudSat cloud type as the target showed a remarkable improvement in elimination of false rain in the precipitation algorithm.

The second approach to identify false rain is to use the satellite radar observation to find location of false rainy pixels. CloudSat is equipped with cloud profiling radar (CPR) that provides radar observations near the surface. The precipitation column algorithm uses the radar data in addition to surface reflection characteristics to identify the occurrence of rain over land. The main advantage of the CloudSat radar compared to ground-based radar is that CloudSat orbits the earth almost at the same time as MODIS. Because precipitation processes can happen in short period of time, having simultaneous observations is an important key to using multiple data sources. An ANN model was trained based on six MODIS channels to make a connection between MODIS observations and rain occurrence. The trained model has the ability to estimate rain or no-rain regions on MODIS imagery.

N. Nasrollahi, *Improving Infrared-Based Precipitation Retrieval Algorithms Using Multi-Spectral Satellite Imagery*, Springer Theses, DOI 10.1007/978-3-319-12081-2_7

In addition to using cloud type classification data from CloudSat, a trained cloud classification model was created in this research using a neural network model to find no-rain clouds on the MODIS image and filter non-precipitating regions. The results show promising outcomes for the summer season data to classify high no-rain clouds with 70 % accuracy. The winter season cloud type classification has some limitation that needs to be further improved.

The following objectives mentioned in Chap. 1 were tested and addressed in this dissertation:

1. Using multi-spectral data in satellite precipitation algorithms will help improve precipitation algorithms. There is a need to move from single IR channel estimations to multi-channel precipitation algorithms. The first objective of this dissertation is to show the effectiveness of using multi-spectral data in satellite precipitation estimation.

To overcome the limitations of IR-based observations in satellite precipitation estimation, multi-spectral data can be used. Moving from single channel to multiple channels in satellite products has been a topic of current precipitation estimation research. Multi-spectral data help to observe some information beyond only top cloud brightness temperature. High vs. low clouds and thin vs. thick clouds are distinguishable when considering brightness temperature in various spectral wavelengths. The distinction between high non-precipitating and deep convective clouds is possible using multi-spectral data. In this research, a set of six MODIS WV and IR channels are used in combination with surface rain occurrence data to find the no-rain regions. Using an ANN model for summer and winter seasons, the performance of more than 77 and 93 % accuracy was achieved for summer and winter seasons, respectively. In addition, the same model was used on real-time PERSIANN precipitation data. Results show false alarm removal of 62 and 61 % for two case studies of summer and winter season PERSIANN data in comparison with ground radar, respectively.

2. The second objective of this dissertation is to show that satellite precipitation algorithms will benefit from information on cloud structure and characteristics. Clouds create precipitation, and adding information about different types of clouds will improve precipitation algorithms.

IR-based algorithms are indirect rainfall estimation techniques that measure the top cloud temperature. The precipitation estimation algorithms use empirical relationships between cloud top temperature and measured rainfall to estimate precipitation. In addition to multi-spectral data, CloudSat provides a unique set of observation of cloud vertical profile that was shown to be an effective tool in satellite-based precipitation estimation. CloudSat cloud type classifies clouds into seven different groups. Among different types of clouds, cirrus and altostratus are non-precipitating. A neural network model was trained to distinguish these cloud types. After identifying non-precipitating cloud coverage regions, the rain estimation from satellite products can be eliminated.

CloudSat only provides cloud type classification on a very narrow swath. In this research, a self organizing feature map (SOFM) model was used to classify clouds into seven types. In this approach, on each MODIS image the clouds are classified into one of seven different types. Finding the non-precipitating clouds are of interest to this study to remove falsely rain pixels. The classification results showed promising results on summertime data.

3. The main reason for false rain observations in satellite-based products is the presence of high cirrus clouds. These highly elevated clouds have cold cloud tops in IR imagery. Therefore, they show false rain signals in satellite-based estimations. The third objective is to show that by identifying and filtering cold cirrus clouds false rain reduces.

A considerable portion of false rain pixels are associated with high cirrus clouds and cirrus anvil. These cloud types with high altitudes are composed of ice crystals. Because of their very cold tops, they often appear as rain in IR-based algorithms. Larger spatial rain coverage on the ground in the case of deep convective storms, confirms the false rain estimation in case of cirrus anvil cloud. Integrating multi-spectral MODIS data and CloudSat observations showed that most of no-rain pixels were associated with high clouds. The trained model was able to identify no-rain high clouds with the accuracy of 91 and 94 % in summer and winter season validation studies, respectively.

7.1 Future Work

The following future research directions are suggested after completing this research.

1. *Deriving various rainfall algorithms based on different cloud types.*
   In this study, a new cloud classification algorithm was developed that showed promising results for summertime cloud type classification. After finding different types of clouds, separate rain estimation algorithms can be developed for each cloud type to achieve a better accuracy in rainfall detection. For example, the type of rainfall events from nimbostratus clouds is prolonged and not very heavy. In contrast, deep convective clouds usually produce intense precipitation events. After finding the type of cloud system, a better estimation of rainfall is foreseeable.
2. *Integrating shortwave infrared channels in the multi-spectral channel consideration.*
   In this study, the spectral data in the range of shortwave IR are not considered due to their solar contamination during daytime. The shortwave data can be used at night or during daytime after correction. In future studies, the shortwave IR data can be added to this study after reflectance correction.

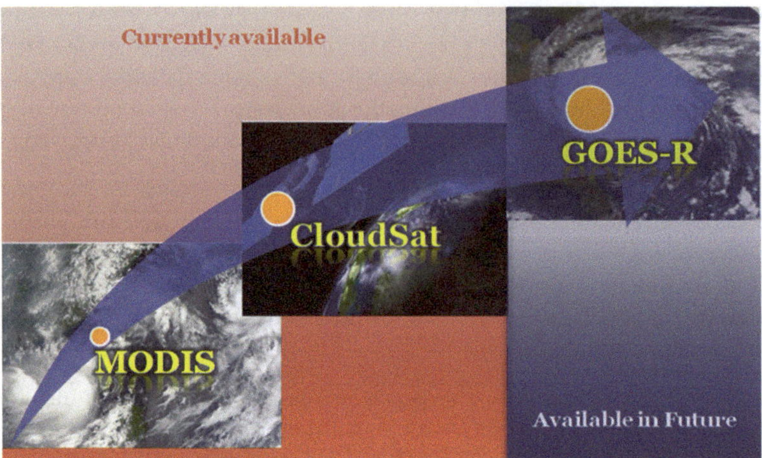

Fig. 7.1 Availability of CloudSat, MODIS and future GOES-R satellites

3. *Adding textural information for cloud classification.*
   In addition to multi-spectral data, textural information is also useful to distinguish different cloud types. The degree of smoothness of the texture is one of the valuable information that can be used to identify homogeneous clouds (e.g. stratiform) vs. non-homogeneous clouds (e.g. convective clouds).
4. *A global cloud classification system is achievable using multi-spectral data available from future GOES-R satellite.*
   In the future, there is a possibility to include multi-spectral data from the Advanced Baseline Imager (ABI) sensor on board the future Geostationary Operational Environmental Satellite-R Series (GOES-R) to overcome the limited retrievals of MODIS. GOES-R is the next generation of geosynchronous environmental satellites was planned to launch in 2015. All the MODIS multi-spectral data that are used in this study will be available from GOES-R and the same trained model is ready to be applied on GOES-R images. Since GOES-R is a geostationary satellite, it will provide higher temporal resolution data available every couple of minutes (Fig. 7.1).